INVEST IN LIVING

WILD FRUITS AND NUTS

D0996054

The *Invest in Living* Series
Getting the Best from Fish
Getting the Best from Meat
Home Goat Keeping
Home Honey Production
Home Made Butter, Cheese and Yoghurt
Home Poultry Keeping
Home Rabbit Keeping
Home Vegetable Production
Meat Preserving at Home

INVEST IN LIVING

WILD FRUITS AND NUTS

by

GEOFFREY ELEY

EP Publishing Limited
1976

Acknowledgements

A good deal of my childhood years was spent with a grandparent and several aunts whose slender resources were augmented by running an art school 'for young ladies' and by the sale of botanical illustrations and flower paintings, a *sine qua non* of which was a close observation of plant and tree life. Without being aware of the open-air expeditions with paints and easel as tuition, I was becoming an amateur naturalist (partly also a very indifferent flower painter!) and I was instilled with an understanding of the English countryside in which I have lived ever since.

So, in writing this book, I acknowledge the perceptiveness and appreciation of all forms of wild life I was given—and, even more important, my thanks are due to the very large numbers of authors whose books on Natural History have been collected and studied since boyhood. It is perhaps fitting that I should therefore also dedicate this modest work to all these naturalists, authors and artists whose distinguished names span four centuries— ranging from and including John Gerard (*Generall Historie of Plantes,* 1597), through the Rev. Gilbert White (*Natural History of Selborne,* 1788) and Richard Jefferies (*Life of the Fields,* 1884) to Geoffrey Grigson—formerly, like myself, a BBC producer—whose recently published work, *The Englishman's Flora*, is a masterly piece of scholarship.

About the author

The author, Dr Geoffrey Eley, MA, was for some years producer of BBC agricultural and horticultural programmes as well as himself being a frequent broadcaster on rural subjects in the Overseas Services. Before devoting himself to book publishing, Dr Eley edited specialist journals concerned with agricultural science and medicine; in 1962 he launched *Medical News* as Britain's first weekly newspaper for doctors. Now an editorial director and consultant to two publishing houses, Dr Eley is a Fellow of the Royal Society of Medicine and a member of the Royal Society of Literature.

ISBN 0 7158 0457 X

Published 1976 by EP Publishing Ltd, East Ardsley, Wakefield, West Yorkshire WF3 2JN

Printed and bound in Brighton England by G. Beard & Son Ltd

Contents

Note on measurements

Many of the recipes in this book are hundreds of years old and have been adapted by successive generations to round imperial measures of pints, ounces and so on. Rather than distort these amounts by converting them to metric measure in the recipes themselves (and also to avoid cluttering the recipes with figures of both standards) we give here tables showing equivalents which can be used when necessary.

oz	grammes
$\frac{1}{4}$	7
$\frac{1}{2}$	14
$\frac{3}{4}$	21
1	28
2	57
3	85
4	113
5	142
6	170
7	198
8	227
9	255
10	283
11	312
12	340
13	369
14	397
15	425
16	454

lb	kg
$\frac{1}{4}$	0.1
$\frac{1}{2}$	0.2
$\frac{3}{4}$	0.3
1	0.4
2	0.9
3	1.4
4	1.8
5	2.3
6	2.7
7	3.2
8	3.6
9	4.1
10	4.5
11	5.0
12	5.4
13	5.9
14	6.4
15	6.8
16	7.3

pints	litres
$\frac{1}{4}$ (5 fl oz)	0.1 (142 ml)
$\frac{1}{2}$ (10 fl oz)	0.3 (284 ml)
$\frac{3}{4}$ (15 fl oz)	0.4 (426 ml)
1 (20 fl oz)	0.6 (568 ml)
2	1.2
3	1.7
4	2.3
5	2.8
6	3.4
7	4.0
8 (one gallon)	4.6

gallons	litres
1	4.6
2	9.1
3	13.6
4	18.2
5	22.7
6	27.3
7	31.8
8	36.4
9	40.9
10	45.5

°F	°C	°F	°C
200	90	350	175
225	110	375	190
250	120	400	205
275	135	425	220
300	150	450	230
325	160		

Introduction

Everyone knows that hazel nuts and blackberries are very nice to eat—but many people are unaware of the abundance of other edible fruits and nuts which can be gathered free in Nature's wild and rich larder.

There are, in fact, food plants, shrubs and trees all over our countryside. They can be found in fields and hedgerows, in woods and at the riverside, even up on the bleak, high moorlands, and searching for them provides an unceasing open-air leisure activity.

This book will enable you to identify the *wholesome* fruits and nuts, besides telling you when they are available and where to look for them. It also includes advice on the best ways to enjoy free wild foods, in many cases with delectable cooking recipes collected during many years from widely different parts of Great Britain—an entertainment in itself.

Wild fruits are given first, starting with the most familiar of all—the blackberry. The second section covers nuts, from that most English of hedgerow small trees, the hazel, to the more exotic walnut. Many other wild foods are, of course, abundant in our countryside and are mainly eaten as leaves, stems or roots.

Every flowering plant produces a fruit, and this contains a seed to ensure survival of the species. Some of these fruits swell and become attractive to some animal or bird so that they eat (and excrete) the seed to germinate farther afield. These tasty fruits are of two kinds and, when you are gathering wild food, you should know the difference.

True succulent fruits develop solely from the ovary of a flower and there are two types among wild plants—the drupe and the berry. The drupe has a single stone or pip in the centre (cherries and plums, for instance). Berries differ from drupes because they have *many* seeds all contained in a single case. Most wild fruits in the British Isles are berries, but some hedgerow plants—the blackberry among them—bear fruits formed from a collection of drupes.

False succulent fruits are those which develop from parts of the flower other than the ovary. In the rose family, the receptacle below the petals, which contains the ovaries, also swells to form the fruit, or rose hip as it is known. Apples and pears also have this particular reproductive structure.

I often get asked—'all very well gathering wild food, but is it safe to eat?' The short answer to this is that there are fortunately relatively few poisonous plants in Britain, and a list of these is given at the back of this book—even then, most of these poisonous plants are uncommon. There is, therefore, little need to be worried on this score when you are out gathering the wild harvest—but you should still make sure, from the descriptions and illustrations given, that you know your trees, shrubs and plants by their growth, leaves, flowers and fruits; also that you realise that the *way* you eat the fruits and nuts is important, since some can make you unwell unless cooked or prepared in special ways according to the recipes given.

Avoid picking wild foods from road verges sprayed with weed killer, from intensive farming areas where insecticides are commonly used, and from main traffic routes where dust and exhaust fumes cover plant life.

So much for your own well-being, but there are factors to take into account. Certainly you, and particularly children if you have them with you, should observe the rules of the 'Country Code'. It is in everyone's interest that gates should be shut, litter taken back home for disposal, and damage of any kind avoided—and this includes hedges and farm crops such as pasture for hay-making, let alone more 'obvious' (to the townsman) fields of grain.

Equally important, do not become a plant vandal. By this I mean that we all must have regard for the vigour and survival of plant life now and for future generations. Do not be greedy and strip a plant of fruit, nuts or leaves— take only a reasonable quantity of what you need and do this without damaging branches or stems of trees or plants. It is socially wrong (and often illegal) to uproot whole wild plants in the hope they will look nice in the garden—and they usually die on you, anyway.

It also helps the visitor to know at least a little of the countryman's law. A sign 'Trespassers will be Prosecuted' often causes disappointment and confusion—needlessly. The fact is that it is an empty threat, meaningless in law unless actual damage is committed. A farmer's warning 'Beware of the Bull'
can be a much more practical matter and one not wisely ignored if the animal is around—he *may* take no notice even if you enter a private field, but again he may angrily resent your presence. (Public footpaths are a different matter: Bye-laws prevent a farmer having adult bulls at large in fields carrying public right of way.)

Trespass is not a crime unless an offence goes with it; when, for instance, entry into private land has been made in search of game or where there is malicious damage to crops.

The blackberries or primroses you gather on private land are natural products of the soil and have no owner until you gather them; the gathering makes you the owner, and the occupier cannot demand the produce, nor can he prosecute the picker for stealing it.

There would, however, be an offence if you pulled up, say, a turnip or sugar beet plant. That would be stealing. Breaking a hedge or fence, and trampling down a growing crop of hay or corn, is damage 'attendant upon the trespass'.

When gathering fruits and nuts, do not pick large quantities from one bush, rather select only the best, dry and insect-free berries from a stretch of hedge and gather the fruits into an open basket, not a paper bag which will probably go soggy and disintegrate before you get home. Gloves, a walking stick, knife and scissors are all useful tools for an expedition after wild foods.

The countryside is overflowing with life. It is worth doing more about it than merely admiring the scenery from your car—and the following pages will tell you just what to gather and where to find it.

WILD FRUITS

Blackberry

All across the world blackberries have always been appreciated. Seeds of the blackberry, or bramble, have even been found in the stomach of a Neolithic man dug up from the Essex clay.

To the townsman blackberries mean more than most other wild fruits. 'Going blackberrying' represents the essence of enjoyment of the countryside, a welcome change from office or factory, a satisfying exercise and complete freedom for the family.

Blackberries are good any way you eat them—raw, cooked in puddings or as jam. A tart made of blackberries and apples, with a nice short crust and a bowl of clotted cream, is a sweet course fit for a king. There are several hundred species and varieties of blackberry (Latin name: *Rubus*), of which two are found in every part of Britain—namely the 'common' blackberry, or bramble, and the dewberry (this carries a smaller fruit than the blackberry, with fewer segments, and is covered with a grey bloom.)

Belonging to the same family is the wild raspberry (*Rubus idaeus*), plentiful on chalk downs.

The fruit of blackberries is made up of a number of drupelets, or juicy seeds. They turn from green through red to a deep purple-black and are ready for picking from August to early October.

Blackberries form in large clusters at the end of older shoots, which die after two or three years' cropping. The berry at the tip of the stalk is the first to ripen as well as being the sweetest and largest—these are the ones you should eat raw and enjoy while you are picking (even if the children eat far more than they put into the basket!). The reason for this is that a little later in the season the other berries ripen but they are less juicy and are therefore best kept for jam making or cooking in pies.

It is interesting to note how blackberries propagate themselves. Each cane starts life growing upright, but then curves downwards until its tip touches the ground, and here the shoot takes root and so gives rise to a clump of new canes.

Blackberry

Easy to Find

The different species of bramble are adapted to living in such varied places as mountainsides and dense woods. Often one of the first plants to arrive on newly cleared ground is the blackberry, its seeds being brought by wind or dropped by passing birds and animals.

The bramble spreads rapidly, creating shelter for other plants as well as for birds and insects. In the spring, the tangle of bramble stems makes an ideal nesting site for many birds, and later in the year the insect life and berries are a major source of food for many British birds. You will sometimes see blackbirds eating the berries, then wiping their bills on the leaves to get rid of the seeds—an action called 'pip-spitting'.

In parts of Great Britain the bramble has been of more significance than being just a fruit, pleasant to eat or for jam making. In the highlands of Scotland a bramble used to be twined with ivy and rowan to ward off witches and evil spirits. It was, said the Highlanders, a blessed bramble with which Christ switched the donkey on the way to Jerusalem—and a piece of bramble with which he drove the moneylenders out of the temple.

Preserving the Fruit

An excellent way of preserving blackberries, keeping them ready to serve as stewed fruit or for making into jam with the addition of extra sugar, is as follows: Boil 2 lb. of sugar with 2 pints of water for 15 minutes; add 8 lb. of blackberries and 1 teaspoonful of salicylic acid, and cook until it is soft but not 'mushy'. Put the product into a stone jar, sprinkle the acid on top, tie down at once with brown paper.

Done this way the blackberries can be used as required and tied down again.

Various recipes for using the blackberries are common enough in the cookery books, but here are some of the less well-known but delightful uses for blackberries:

Spiced Bramble Jelly

Put washed blackberries into a preserving pan and cover with water. Add nutmeg, mace and cinnamon in the proportion of a saltspoonful of mixed spices to each pound of fruit. Bring to the boil and then simmer very slowly until all the juice is extracted from the fruit. Then pour through a jelly bag. Next day, put the juice in the pan with 1 lb. of sugar to each pint of juice and boil rapidly for 30 minutes—or until it jellies. Put into warm jars and tie down.

With the pulp left over in the jelly bag, you can make delicious spiced bramble 'cheese'—and avoid all waste at the same time.

Rub your pulp through a wire sieve and to each pound add $\frac{3}{4}$ lb. of sugar and leave for some hours to dissolve. Add the juice and grated rind of a lemon; boil all together, stirring carefully all the time. Pot and seal. A splendid preserve for breakfast!

A Wine like Port

To make blackberry wine all you need to do is place alternate layers of ripe blackberries and sugar in wide-mouthed jars and allow to stand for three weeks. Then strain off the liquid and bottle, adding a couple of raisins to each bottle. Cork lightly at first and later more tightly.

Not only is this wine cheap, but it has a flavour like that of good Port and it will keep in prime condition for at least a year.

Blackberry and Apple Marmalade

Boil together equal quantities of apples and blackberries. Do not core or peel the apples but cut them up roughly, add the brambles and barely cover with cold water. Bring very slowly to the boil, and when the apples are reduced to a pulp, rub the contents of the pan through a sieve—leaving only the skins and seeds.

To every pound of pulp add 1 lb. of sugar. Boil together for half an hour—stirring constantly otherwise the pulp may burn.

If poured into moulds this marmalade will turn out quite stiffly and (according to an old Aberdeenshire recipe) is delicious served with milk pudding.

From Devonshire comes a recipe for—

Bramble and Sloe Jelly

With this one the trick is to boil the fruit well until bleached-looking—and you need 8 lb. of blackberries and 2 lb. of sloes. After boiling, squeeze through double butter muslin, or jelly bag. To 7 or 8 pints of the juice add 5 lb. of sugar and boil up as quickly as possible—half an hour or less. These quantities make nearly 12 lb. of good, clear jelly.

Blackberry Cordial

An old Essex recipe gives us an excellent winter remedy for colds and sore throats. All you want is a table-spoonful of blackberry cordial in a glass of hot water at bed time—and this is how to make it:

Pour a pint of white wine vinegar over a quart measure of ripe blackberries. Let this stand in an earthenware jar for a week, stirring meanwhile to extract the juices. When ready, strain off and put the liquor in an enamel saucepan with 1 lb. of cubed sugar and $\frac{1}{2}$ lb. of honey. Bring to the boil, then remove from the heat and allow to get cold. Bottle and cork and keep in a dark place.

Bramble Junket

Perhaps the most delicate of all blackberry recipes is the very old one for bramble junket—old, yes, but not very well known.

Take a square of coarse, strong cheesecloth and pile it full of the ripest blackberries. Knot the four corners, slip a stick under them and twist over a china bowl, pressing the bag with a wooden spoon until you have a bowl full of rich, thick blackberry. Do not add anything to it, and it should set solid in about two hours.

This product will be of the consistency of junket. Eat the junket with cream and thin brown bread and butter, sponge fingers or biscuits.

Other Uses

During the Industrial Revolution blackberries were put into buns instead of currants; poor people cooked parsnips and beetroot using blackberries to sweeten them.

Blackberries used to be collected by country people and sold for dyes;

navy blue and indigo were originally blackberry juice. It was also used to dye woollen stockings black and even to re-dip black silk that had gone green (mixed with ivy leaves it gave a heavy black). There are also 17th- and 18th-century references to maids wearing 'pale lavender ribbons', and this delicate tint was obtained from bramble dye.

Another custom was to pick the tips of growing leaves, dry them and then rub down to a powder to make 'black tea'. A trade in old tea leaves (when real tea was very nearly £1 for 1 lb.) was created by adding old tea leaves, re-dried, to blackberry leaves.

Local Names

All plants have localised names, and frequently very pleasing and descriptive they are. Here are some for the blackberry:

Blackblegs and blaggs, Yorkshire; black kites, Cumberland; brammel kites, Durham; bumble kites, Hampshire and north-western districts; doctor's medicine, Somerset; garten berries, Scotland; and mushes, Devon.

Elderberry

The elder is a very widespread and common tree of the wayside, hedge-rows and waste places. None of our trees grows more rapidly in its earliest years, and any bit of its living wood easily takes root. The elder has a corky bark, a scaly surface to the young twigs and five large dark green and slightly toothed leaves. In June and July the elder is covered with hundreds of tiny cream-coloured flowers.

It is sometimes said that the sweet-smelling flowers of the elder have more uses for mankind than any other blossom. In medicine they have an acknowledged place as an ingredient in skin ointments and eye lotions, and there is evidence of the use of elder-berries in Egyptian medicine.

John Evelyn, the 17th-century diarist, in his celebrated work, *Sylva, or a Discourse of Forest Trees* (1664), praises the elder like this—'If the medicinal properties of the leaves, bark, berries, etc. were thoroughly known, I cannot tell what our country-man could ail for which he might not fetch a remedy from every hedge, either for sickness or wound.'

The juicy shoots of the elder soon harden into a tube of tough wood with a core of pith which is easily scraped out, to the delight of small boys who make peashooters from the tubes.

The ease with which the pith can be cleaned out has given the tree its northern names of bourtree, bore-tree, or bottery; and indeed even the name elder has its relation to the tubular shoots. Anglo-Saxon words indicate that the elder shoots were used like a hollow bamboo in the tropics to blow up a fire. It is known that the housewife got her bellows, the musician his pipe and the schoolboy his popgun—all from the elder.

Folklore surrounds the elder in all parts of the world.

Elder must not be burned. In Warwickshire, for instance, if you put it on the fire you will see the Devil come down the chimney—and in neighbouring Derbyshire a common name for elder is Devil's wood.

The Irish believe it is foolish to make a baby's cradle from elder wood—the child sickens, they say, and the fairies would steal it.

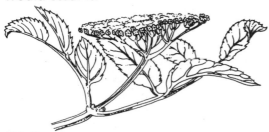
Elderflowers

The elder (Latin name: *Sambucus nigra*) is the only tree which rabbits find distasteful, hence its so often being found on rabbit warrens. It rarely grows more than about 6 m (20 ft.) high and its timber is useless.

The elder bears its fruits between August and October in clusters of small, reddish-black berries. These berries are ripe for picking when the clusters begin to turn upside down. Gather the clusters whole by cutting them with about two inches of stem attached, picking only those where the berries are free from insect pests, and have not started to wrinkle or melt.

For use with any of the recipes to follow it is best to wash the elderberries well and strip them from the stalks with a fork.

Elderberries

Elderberry Wine

This is probably one of the favourite uses for the fruits of the elder. A recipe from Devonshire recommends using 7 lb. of berries and 2 gallons of water. To each gallon of liquid add 3 lb. of loaf sugar, 1 lb. of raisins, half an ounce of ground ginger, half an ounce of whole ginger, bruised, six cloves, half a stick of cinnamon and a lemon.

Strip the berries from the stalks and pour boiling water over them. Let them stand for 24 hours then bruise them well and strain through a sieve or jelly bag. Measure the liquid, put it into an earthenware pan and add sugar and lemon cut in slices. Next, boil the cloves, ginger, raisins and cinnamon in a little of the liquid. Strain and add to the rest of the wine.

Allow to stand for a few days, then take off the cap. Strain again and pour into stone jars or casks. Leave open for a few weeks, continually adding more wine until fermentation ceases. Bung tightly and let it remain for six months before bottling.

Elder Syrup

Elderberries have tonic and health-giving qualities. Years ago the elderberry's power in magic was equalled only by its power in medicine; it was considered by medieval physicians to be a remedy 'against all infirmities whatever'. True or otherwise, the following recipe certainly makes a first-rate syrup which, drunk hot, is helpful against winter coughs and colds:

Into half a gallon of elderberry juice add the white of an egg, well beaten to a froth, and put this in a pan over a slow fire.

When it begins to boil, skim it as long as any froth rises; then put to each pint 1 lb. of sugar and boil slowly until you have a good syrup—you can tell this by testing a little on your nail and, if it congeals, then it is done sufficiently.

Let the syrup stand and when it is cool put into bottles covered on the neck with paper pricked full of holes.

Elderberry Pickle

A Northumbrian recipe advises the use of 1 lb. of elderberries washed, stalks removed and well mashed. To the elderberries add two tablespoonsful of sugar, a small onion, half a teaspoonful of ground ginger and the same quantity of ground mixed spice, half a pint of vinegar and just a pinch of salt.

Put all these ingredients into an enamelled pan and bring to the boil. Cook slowly until the mixture is thick, stirring so that it does not burn. Bottle while hot and tie down.

Elderberry Sauce

From the hunting country of the Midlands, and Leicestershire in particular, comes this aristocratic recipe:

Pour one pint of boiling claret over one pint of elderberries in a stone jar or casserole dish. Cover and place overnight in an oven at a very low temperature. The following day the liquid can be poured into a saucepan with one teaspoonful of salt, a few peppercorns, and clover, a finely chopped onion and a piece of root ginger, bruised. Boil the lot for 10 minutes and then bottle securely with the spices.

According to one authority, Richard Mabey, the fruity-tasting elderberry sauce used to be kept for as long as seven years, and is especially recommended for use with liver.

Elderberry Jelly

Add one pint of water and six tablespoonsful of lemon juice (*or* two teaspoonsful of tartaric or citric acid will do) to 4 lb. of fruit and simmer for between 40 and 60 minutes. Strain the elderberry juice through a jelly bag, measure, and to each pint add $\frac{3}{4}$ lb. of sugar.

Boil until the jelly sets, stirring meanwhile. Pectin can be added to the jelly to speed setting.

———

All the above recipes have been for the *berries* of the elder plant—but there are some delicious ways of using the elder *flowers*, the best known recipes being for elderflower wine. It is not generally known, but the aroma of elder flowers is so strong that a bunch of the flowers, drawn through any fine jam just before bottling, will scent the jam deliciously. And, incidentally, elder flowers are very tasty eaten straight off the tree on a hot summer's day.

Elderflower Wine

This is a clear, sparkling wine and as good a recipe as any is the following one given me by a Cambridgeshire woman when I lived in that county:

Dissolve 3 lb. of sugar in a gallon of water, stir in the well-beaten white of egg, bring to the boil and keep it so for 30 minutes, then skim. Have ready in an earthenware bowl 2 breakfast cupsful of loosely packed elder flowers (stripped, of course, from their stems), 1 lb. of raisins (split but not stoned) and the juice of a large lemon. Pour them over the water mixture. Stir thoroughly, and when little more than raw add 1 oz. of yeast.

For the next 10 days you must stir once a day, then strain and put into a clean, dry stone jar, setting the cork in only loosely. When the mixture has done working, cork tightly—and in six months' time it will be ready for bottling. Take care to strain the liquid through flannel and ensure that your bottles and corks are thoroughly sterilised and dry to receive the wine.

A little elderflower wine makes a pleasing flavouring essence for cakes, junkets and light sweets; it can also be used as a money-saving substitute for lemon juice in icing.

From Austria

Try elderflower pancakes, a great favourite with the country people in parts of Austria. All you have to do is to hold circlets of blossom by the stem, dip into batter and fry. Eaten with sugar these elderflower pancakes are both delicious and fragrant.

Elderflower and Rhubarb Jam

Cut 4 lb. of freshly pulled rhubarb into one-inch lengths. Put the cut rhubarb in a preserving pan with a large cupful of water and the juice of two lemons. Cook gently until the rhubarb is soft—then weigh and to every pound add 12 oz. of sugar, stirring until the sugar dissolves.

Suspend from the handle of the preserving pan a muslin bag containing three sprays of freshly-gathered elderflowers; boil for 20 minutes, stirring gently. Test for setting and if necessary boil for another five minutes. When ready, put immediately into hot jars and seal at once.

Elderflower Ointment

This is an old-fashioned medical use for the flowers of the elder—said to be good for heat lumps, bites, chapped or rough hands, and excellent for a baby in case of rash or roughness of the skin. 'Equally suitable for young and old', says an old Staffordshire recipe. Here it is:

Strip 1 lb. of elderflowers from the stalks. Put them in a saucepan with $\frac{1}{2}$ lb. of unsalted lard. Simmer very gently till it turns a pale green, pressing the flowers and stirring often. Then strain through muslin and pour into small jars; leave to set and cover.

The elderflower ointment is ready for use as soon as it is cold—and, apart from the medical uses already listed, it is said to be splendid for lady gardeners' hands—and certainly much cheaper than concoctions you buy at the chemist's shop.

Local Names

Bourtree is a common name for elder in the northern counties; Devil's wood in Derbyshire; dog-tree in Yorkshire; and in Dorset the oddest local name of all—God's stinking tree.

Crab Apple

The wild apple is found all over the United Kingdom as far north as the Clyde, and it has a long history—crab apple pips have been found in the remains of prehistoric cooking pots.

Crab apple

The crab apple is a small tree with reddish-brown twigs and leaves which are oval in shape, toothed and usually downy. The tree is common in our hedgerows, in woods and on heaths.

Wild apples are round, yellowish-green, sometimes turning scarlet, and they can be picked any time after July —but before autumn gales have blown the fruit away.

Because the crab apple must have sufficient light to do well, the best fruit is usually found on trees in hedges and open heaths. Not only is the wild apple a very variable species, but many cultivated apples have seeded themselves in the wild and either reverted to a wild form or have become crossed with a true crab apple tree. This is one of the reasons for such

a wide variety of this wild fruit being found.

In the typical form of the wild apple (Latin name: *Malus sylvestris*) the yellow and red fruit hang by their slender stalks, but there is a variety *(Pyrus mitis)* in which the fruit is borne above the stouter stalks. This latter variety may also be recognised by the downiness of the young leaves.

Crab apples are normally too acid to eat raw, though children cannot resist them even if it means stomach-ache afterwards! However, due to the large number of varieties, one can occasionally find a crab apple sweet enough to eat raw.

Crab apple derives its name from the old Norse word, skrab, meaning a small rough tree. The delicately pink and white perfumed flowers open late in April above the leaves. The crab apple seldom grows higher than about 9 m (30 ft.) and usually lives for only about 50 years.

Like garden apples, the wild apple grows by putting out long shoots, but flowers and fruit come only on short shoots, or spurs, as they are called by fruit growers. The flowers are typical of the rose family, to which the crab apple belongs.

Birds eating the apples scatter the ripe, black pips and this is the reason seedlings—and later wild apple trees— are so often found in hedgerows where birds roost.

There are, too, beautiful cultivated

varieties of crab apple with autumn fruits richly coloured in red, rose and gold—and these are used for making crab apple jelly. The wild apple is the origin of all our dessert apples, and of these there are over 3,000 named varieties.

Crab Apple Jelly

Wash thoroughly 6 lb. of crab apples and cut them in halves. Place them together with five pints of water and boil for half an hour. By that time the fruit should be soft and broken. Strain through a large colander, then put the liquid in a jelly bag suspended over a big basin. (An old pair of women's tights makes an entirely acceptable substitute for a proper jelly-bag.)

Next day, measure the crab apple juice and pour into a pan—with some flavouring tied to the handle of the pan in a piece of muslin—and, after boiling for half an hour, add sugar at the rate of 1 lb. per pint of juice and remove the muslin bag. When the sugar has dissolved bring the liquid to the boil again and boil quickly for 15 minutes. Test for setting and when ready pot up —making sure you remove every particle of scum.

Crab apples make a beautiful jelly, rich glowing pink, but the juice lacks flavour—hence the advice above to add some flavouring. This can be three or four leaves of the old oak-leaved scented geranium—one leaf per pint is enough.

Other delicious flavourings, recommended by such experts as Mrs Arthur Webb and which have 'stood the test of time', can be obtained by using two or three peach leaves in the muslin bag suspended in the jelly whilst cooking; or one leaf of lemon verbena per pint

of liquid if you have this plant in the garden. Again, five or six mint leaves will impart their own pungent flavour to the crab apple jelly; and the petals of freshly-gathered red roses are said to give an elusive but pleasant taste.

A little known but excellent tip for the safe keeping of crab apple jelly: After filling your jars to within 3 mm ($\frac{1}{8}$ in.) of the rim, brush the surface of the jelly lightly with glycerine—then press a waxed disc on smoothly and follow with a suitable cover well fastened. Keep in complete darkness but, like jam, free from changing temperatures and on well-ventilated shelves.

A Pleasant Wine

Crab apples, allowed to ferment, make a pleasant wine and one which improves with keeping. The flavour resembles a sweet Graves.

To make crab apple wine put a gallon measure of sliced crab apples into a gallon of water and let them soak for a fortnight. Strain, and add 3 lb. of brown sugar to each gallon of liquor. Stir well and frequently until fermentation takes place, usually in a day and a half.

The Lancashire recipe quoted here goes on to say that the wine should be left for three days and then put into casks or jars. Lay muslin over the opening until the hissing noise (which tells that the wine is working) has ceased. Then cork tightly, and bottle after three months.

The wine should be matured in a cool, dark place and later decanted into fresh bottles, tightly corked. Properly done, crab apple wine is one of the very best country wines.

Shakespearian

'Lambs' wool' is the name given to an historic crab apple recipe. It is a warming drink made from hot ale, spices and crab apples—the drink Shakespeare refers to when he talks of 'roasted crabs hissing in the bowl' *(Love's Labour's Lost)*.

Puck, too, mentions the drink in *Midsummer Night's Dream*:

*And sometimes lurk I in a gossip's
   bowl,
In very likeness of a roasted crab,*

*And when she drinks, against her lips
   I bob.*

Like the quince, two or three wild crab apples improve an apple tart.

Local Names

Scrab, North Country; gribble, Dorset and Somerset; grindstone apple, Wiltshire; scrogg (i.e. a scrubby tree or bush), Durham and Border Country; wilding tree, Shropshire.

Wild Rose

The wild or dog rose (Latin name: *Rosa canina*) is England's national flower.

Over 100 species of wild rose occur in Britain; besides the dog rose, other commonly found varieties are the downy rose, burnet rose, field rose and apple-scented rose.

Commonest of all is the dog rose, a bush abundant in hedgerows everywhere except in Scotland, and showing its familiar shell-pink flowers in June and July. This is the rose cultivated by nurserymen to provide rootstocks on which the bigger garden varieties of rose are grafted.

Scotland's most frequently found species is the downy rose *(Rosa villosa)*. In contrast to the dog rose, which grows up to 3 m (10 ft.) high and has hairless leaves, the downy rose is a low shrub, 0.9 to 1.8 m (3 to 6 ft.) high, with leaves densely covered in soft hairs. The downy rose has deep pink flowers between June and August, but it is not often found in southern England.

The burnet rose *(Rosa spinosissima)* is a low-growing shrub with small leaves—much like those of the salad burnet (a common plant on chalk grassland in England and Wales, bearing balls of green flowers between May and June and whose crushed leaves smell like cucumber). Flowers of the burnet rose are small, solitary and creamy-white in colour. These can be found, especially in sand dunes near the sea, from May to July, and the blossoms develop in the autumn into purple-black fruits on hips (the only wild rose hips that are not scarlet or red).

Next, the field rose *(Rosa arvensis)* which—despite its 'field' name—is usually seen in woods and deeply shaded places. This rose also grows over hedges, sometimes climbing up to 1.8 m (6 ft.) where there is enough support for its flexible stems. Clusters of pure white flowers open in June and July and these are followed in the autumn by small, bright red hips. This is another species, like the dog rose, absent from Scotland.

The apple-scented rose *(Rosa rubiginosa),* or sweet briar, is usually smelt before it is seen—especially on a warm, showery day. The apple fragrance is given off by brown-coloured glands on the undersides of the leaves. This branching shrub grows up to 1.8 m (6 ft.) high and the June and July flowers of the apple-scented rose are bright pink.

The fruits or hips of wild roses are rich in vitamin C. During the last war civilians and their families were asked to collect the hips each autumn, and as a result no less than $2\frac{1}{2}$ million bottles of home-grown rose hip syrup were produced—equivalent in vitamin content to 25 million imported oranges.

Rose hip syrup manufacturers still rely on voluntary collectors—mostly schoolchildren, who receive a few pence per pound as an incentive, and enormous amounts of these wild fruits are collected every year, particularly in the North. Many schools collect as much as 450 kg (1,000 lb.) in a winter, and a few years ago one Carlisle secondary school exceeded 13,600 kg (30,000 lb.) in one year.

Certainly, the gathering of rose hips is a dramatic example of the value of wild food use—but it is by no means

Wild rose

a modern development. When cultivated fruit was scarce in the Middle Ages, rose hips were used as a dessert. Gerard, the 16th-century herbalist, says of rose hips—'when it is ripe, the fruit maketh the most pleasant meates and banketting dishes or tartes'.

Although rose hip syrup—vitamins complete—was in recorded use as early as 1730, it was not until 1934 that it was scientifically found that wild rose hips contain more vitamin C than any other fruit or vegetable— four times as much as blackcurrants and *20 times* as much as oranges!

The syrup is the basis of rose hip recipes, and making it is the best way of getting rid of the troublesome hairs prickles on this fruit. Years ago the Ministry of Food issued a *Hedgerow Harvest* leaflet which included the following sound advice:

'Have ready 3 pints of boiling water, mince the hips in a coarse mincer, drop immediately into the boiling water or if possible mince the hips directly into the boiling water and again bring to the boil. Stop heating and place aside for 15 minutes. Pour into a flannel or linen crash jelly bag and allow to drip until the bulk of the liquid has come through. Return the residue to the saucepan, add $1\frac{1}{2}$ pints of boiling water, stir and allow to stand for 10 minutes. Pour back into the jelly bag and allow to drip.

'To make sure all the sharp hairs are removed put back the first half cupful of liquid and allow to drip through again. Put the mixed juice into a clean saucepan and boil down until the juice measures about $1\frac{1}{2}$ pints, then add $1\frac{1}{4}$ lb. of sugar and boil for a further five minutes.

'Pour into hot sterile bottles and seal at once. If corks are used these should have been boiled for $\frac{1}{4}$ hour just previously and after insertion coated with melted paraffin wax.

'It is advisable to use small bottles as the syrup will not keep for more than a week or two once the bottle is opened. Store in a dark cupboard.'

Besides being used diluted as a drink, rose hip syrup is good as a flavouring for milk puddings, ice-cream or most sweets.

Rose Hip Wine

A method of making this lovely drink (much used by grandmother in my native county of Warwickshire!) is to pick a gallon measure of hips and to these add 2 oz. of ginger, two sliced oranges, two sliced lemons, and a gallon of water. Boil until the hips are tender and allow to stand for three days, stirring daily.

Next, strain and add to each gallon of liquor $2\frac{1}{2}$ to 3 lb. of sugar, the juice of one lemon and $\frac{1}{2}$ oz. of yeast. Cask and, after three days, bung up.

The rose hip wine can be drunk after three months—or bottled up and kept for another three to nine months.

Rose Hip Cream

For this recipe I have to thank a Yorkshire farmer's wife. And this is how she makes the cream for serving with custard, or milk, in small fruit plates:

Pick only the larger, red and entirely ripe hips. With a knife, split the hips open and, using about $\frac{1}{2}$ pint of hips from which all the seeds and hairs have been removed, put them in 2 pints of water and bring slowly to the boil.

Allow your fruit to boil until the hips are soft, occasionally stirring briskly. Then pass the mixture through a fine sieve and boil up the juice with $\frac{1}{4}$ lb. of sugar and a small piece of cinnamon. When it has boiled, remove any scum floating on the top.

Next, mix up to a thin cream three tablespoonsful of cornflour in a little cold water and pour this into the hot rose hip syrup mixture. Boil again for a short time and pour into jelly moulds or a basin.

This cream can be made during the winter too, merely by drying the hips (after splitting open and removing the seeds) in a cool oven or in the sun and storing away in glass jars for later use.

Rose Hip Apple Jelly

This Argyllshire recipe requires 4 lb. of windfall apples and 2 lb. of just ripe rose hips. It makes a most attractive rose-coloured jelly of exquisite flavour.

Cut up the apples and put into a preserving pan with enough water to cover, plus one pint extra for rose hips. While the apples are cooling to a pulp put rose hips through the coarsest cutter of the mincer.

Add the minced rose hips when the apples are cooked and simmer for 10 minutes. Leave for a further 10 minutes before straining through a jelly bag. Leave to drip overnight.

Next day measure the juice and allow 14 oz. of sugar to each pint of juice. Measure the sugar and heat it thoroughly in the oven before adding to the juice when this has been boiled for three minutes. Finally, test for setting and when ready pour into warm jars and tie down.

A similar recipe from Somerset calls the preserve rose hip honey.

It must be admitted that cleaning out the seeds and hairs from rose hips is a demanding task, needing much patience. But it must be done, for jam as well, since if any hairs are left on the skins in a preserve they can be a very unpleasant internal irritant.

Apart from the hips, the five-petalled flowers of wild roses have their uses, culinary and otherwise (religious rosaries originally consisted of strings of beads made from pressed petals, and the Romans adorned their tables with wild rose petals).

The wild rose scarcely ever droops before it sheds its petals and the right time to gather petals is towards the end of July when the petals begin to drop—pick those that remain into a basket (do not pick and damage young flowers before they have reached the petal-fall stage).

Rose Petal Jam

This delicious—and very sweet—preserve is popular in the Middle East, especially with yoghurt. But from home territory comes this recipe (Sussex) for which wild rose petals or those of the old-fashioned garden damask rose can be used (although these are more difficult to reduce to jelly):

As you accumulate rose petals, perhaps over three or four days, put them into a deep earthenware bowl, squeezing a little lemon juice over the petals. Then cover the crock.

When you have enough petals, measure by weighing and allow $\frac{1}{2}$ lb. of sugar and a similar quantity of honey to each pound of petals. Add a very little water, then boil gently until the jam sets. Bottle in the usual way.

Local Names

These include, for the fruit of the dog rose, cankers in East Anglia; cat-jugs or choops, Yorkshire and Durham; haws, Dorset; hedgy-pedgies, Wiltshire; hips—general name used in many parts of the country; hipsons, Oxfordshire; itching berries (probably because children put the seeds down each other's necks to produce itching), Lancashire; pixie pears, Devon and Hampshire; and soldiers, in Kent.

The local names given the plant itself include briar or (northern) breer; brimmle, Shropshire; canker-rose, Devon, Somerset and Kent; cat rose, Cheshire; can-whin, Yorkshire and

Northern England; cock-bramble, Suffolk; neddy-grinnel, Worcestershire; and pig-rose in Devon and Cornwall.

Local names for the burnet rose are cat-hep in the North; cat-rose and cat-whin in Yorkshire and Northumberland; St. David's rose in South Wales; and the fox-rose in Warwickshire.

Sloe (blackthorn)

With the single exception of the hazel, all our native fruit trees are—like the crab apple mentioned earlier—members of the extensive Rose family. Before the Roman invasion brought improved and cultivated varieties, our forefathers were glad enough to eat raw the sloes, crab apples and wild cherries that we now regard as too sour except when they are cooked.

Sloes

The sloe or blackthorn is a rigid and many-branched shrubby tree up to 3.5 m (12 ft.) high. It has stiletto-like spines and is easily found on most commons and in many hedgerows. The blackish bark that gives its name to the shrub sets off very well the white starry blossoms that come out in the coldest weather of March or April and well before the leaves themselves appear. 'Many sloes', an old saying has it, 'many cold toes'. In autumn the leaf colours are gold, crimson and purple.

Although the sloe is the sourest berry you will ever taste, it is the ancestor of all our delicious cultivated plums. Because the branches and twigs turn in every direction, it is impossible to pick sloe berries without being badly scratched by the thorns; so be prepared and wear gloves. The best time for picking sloes, which mature in October, is immediately after the first frost—this makes the skins softer and more permeable.

The fruit of the blackthorn varies, according to season, from marble-size up to that of a damson. It is round and held erect on a short stalk. It is black in colour, but its blackness is hidden by a delicate 'bloom' that gives the sloe a purplish hue.

One bite of an unripe sloe will set your teeth on edge. However, when the sloe skin begins to pucker, the juice condenses into a more mealy flesh. One may then eat the berries slowly, enjoying the piquancy of each before swallowing. Sloes are notable for making into wine and some authorities have said that some wine sold as Port began with the skins of British sloes instead of Portuguese grapes. An old jibe against the country grocer was that most of his China tea had been grown on local blackthorn bushes! Certainly blackthorn leaves have been made into tea, and the blossoms and juice of sloes have long been used in country medicine.

'For the stomach's sake' was an old-wives' excuse for drinking gin in which sloes have been soaked for some months, and John Gerard, in his *Herbal* of 1597 said:

The juice of sloes du suck the belly, the lashe and bloody flixie, the inordinate course of women's termes, and all other issues of blood in man or woman, and may very well be used instead of acatia, which is a thornie tree growing in Egypt, very hard to be gotten, and of a deere price, and therefore the better for wantons; albeit our Plums of this countrie are equall unto it in vertues.

Lots of legends have grown up around the sloe—and one of these is that it is unlucky to bring the blossom (often called whitethorn) into the house. However, the strong shoots make good walking sticks, and in Ireland rough pieces of blackthorn are used to make a shillelagh (cudgel secured to the wrist by a strap).

Bullace and Wild Plum

The Latin name for the sloe is *Prunus spinosa,* literally meaning 'spiny plum' —and two sub-species are the bullace and the wild plum.

The bullace differs from the sloe in having brown instead of black bark. The branches are straight and, unlike the blackthorn, only a few of them end in spines. Both the flowers and the leaves are broader than those of the sloe and the egg-shaped fruit can be either black or yellow.

The bullace was the domestic plum of years ago, before better dessert plums had been cultivated.

The wild plum also has brown bark and straight branches not ending in spines. The downiness on the underside of bullace leaves is, in the case of the wild plum, restricted to the ribs of the leaf. The fruit is much bigger than that of a sloe and the tree is only occasionally found in hedgerows.

Sloe Gin

Much the best use for sloes is preserving them in gin.

After the first frost (which burns the skin texture slightly making it more easily penetrable) gather about a pound of sloes direct into a fruit-bottling jar, and fill with gin. Cork securely and place in a warm but dark cupboard. By Christmas time your sloe gin, with its piquant flavour, may be poured off and used, but can stand longer with advantage.

If the berries have not been through a frost, pierce the skins with a skewer before putting into the gin.

Sloe gin can be used in cocktails. In earlier days, blended with the small creeping plant commonly called pennyroyal (*Mentha pulegium*) and with valerian, it was used by country wives in the belief that it would cause them to abort—hence the name Mother's Ruin.

Some recipes for sloe gin include using an equal weight of sugar. If you do include this you should, during the storage period, occasionally shake the bottles to disturb the sugar.

Sloe gin is a deep pink-coloured liqueur, quite potent, sour-sweet to taste—and very refreshing. You can eat the berries from the bottle which, by the time you pour off the liquid, will have lost their sourness—to say nothing of it being a shame to waste them as they will have soaked up quite a bit of gin themselves!

Sloe Jelly

This is an unusual, little-known jelly recipe which (like several others in this book) I 'inherited' from a farming aunt in Warwickshire. Sloe jelly, when carefully made, is especially good with game, roast mutton, veal and venison.

Gather 6 lb. of sloes after the fruit has had two or three early frosts on it. Wash the sloes and remove bits of leaves, stems and suchlike before pricking some of the fruit with a darning needle and putting them in a preserving pan with 4 pt. of water.

Simmer for two hours, turn into a jelly bag and leave to drain all night. Next day bring the juice to the boil before adding $1\frac{1}{2}$ lb. of sugar and the juice of four oranges. Stir until the sugar has dissolved, then boil up again briskly for 15 minutes before testing. When the sloe jelly is ready to set, pot up quickly into hot small jars.

The following two recipes are from Ireland:

Sloe and Apple Jelly

Stew equal quantities of ripe sloes and green apples (skins and cores included) until soft, barely covering the fruit in the stew-pan with water. Strain through a jelly bag, and to each pint of juice add 1 lb. of sugar. Bring to the boil, and boil until a little sets when tested.

This sharp, pungent jelly is splendid with mutton, hare or rabbit.

Sloe Cheese

Make exactly as the recipe for sloe and apple jelly but put the cooked fruit through a sieve instead of a jelly bag. Add 14 oz. of sugar to each pint of pulp.

Local Names

There are a number of localised names both for the shrub and for its fruit.

In the north of England the fruit is called slaa; in the West Country it is heg-pegs, hedge-speaks and snags. The plant names include buckthorn, Lincolnshire; bullister, Cumberland, Scotland and Ireland; scrogg, Nottinghamshire and the north country; pig-in-the-hedge, Hampshire.

Hawthorn

The hawthorn hedges dividing the fields in every part of lowland Britain are a man-made feature of our landscape. Hawthorn was planted, mainly in the 18th century, during the great land enclosure movement and as a cheap barrier to prevent farm livestock from straying.

Hawthorn also springs up frequently as a bird-sown bush reaching over 6 m (20 ft.) high. Apart from its use as hedging, the hawthorn colonises waste land everywhere. The bush shelters smaller plants from the weather, and its thorns give protection against grazing animals. So the seedlings of trees like oak and ash grow in safety beneath the hawthorn and eventually supplant it.

The Latin name of the common bush is *Crataegus monogyna*—it has only one hard seed in each fruit or haw; in the Midlands, however, the hawthorn species *(Crataegus oxyacanthoides)* has two or three seeds in each haw. The name hawthorn is Anglo-Saxon (haegthorn) meaning hedge-thorn, and it is similar in German, Dutch, Swedish, Danish and Norwegian. The shrub is also called whitethorn—to make the distinction between its light grey bark and the blackish bark of the blackthorn —and May because it is in that month that the white (and sometimes pink), scented blossom attracts most attention.

The lobed leaves of the hawthorn are very variable, both in size and shape. They are a favourite food of horses and cattle, who would demolish the hedges that confine them to the fields but for the hawthorn's sharp spines. In April, the fresh, bright green leaf buds are called 'bread and cheese' (or, in Wales, 'Barra cause') and are often the first wild vegetable eaten by country children.

'Bread and Cheese'

At this stage the young leaves have a nutty taste and make an excellent addition to potato and beetroot salads (one teacupful of hawthorn buds to four times this quantity of potato salad).

In May, the blossom comes in clusters on unclipped trees, never on close-trimmed hedges however tidy and efficient these are on the farms. Each flower has five petals and by October the familiar red haws attract many birds, particularly finches, tits and members of the thrush family. The haws, which we can use for food, are also an important part of the diet of the woodmouse and other little animals.

Hawthorn is long-lived, sometimes as long as 200 years. Its wood is both hard and tough and its botanical name is derived from the Greek *kratos*, meaning strength. Very beautiful red- and pink-flowered varieties, both single- and double-petalled, are grown as garden trees.

May Day is a festival of vegetation, sex and fertility—and the bringing in of summer. The customs of the May-pole, the May Queen and the dancing are part of our French inheritance. It used to be a May Day 'fertility' custom in France to place branches of haw-thorn outside the windows of young

May blossom

girls and the exciting, musky scent of May blossom (which comes partly from trimethylamine, a volatile chemical which the flowers contain) was thought to make the blossoms suggestive of sex.

The 'wicked' suggestiveness of May Day was hated by the more fervent Puritans of the 16th and 17th centuries and before that—in the early 14th century—the Church attempted to sanctify hawthorn rather than oppose its sexual and fertility connotations. Hawthorn, as carved foliage, entered the churches—a good example is in the chapter house of Southwell Minster in Nottinghamshire, where heads wreathed in hawthorn are portrayed along with other May Day magical

plants, including oak, ivy, cinquefoil, buttercup and maple.

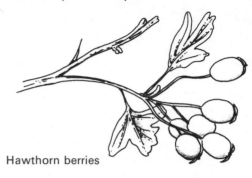

Hawthorn berries

Hawthorn was supernaturally power-ful against many evils, particularly witches. In Ireland, lone hawthorns belong to the fairies—thus the Victorian poet, William Allingham, in his poem *The Fairies* ('Up the airy mountain, down the rushy glen,' etc.) declares:

By the craggy hillside
Through the mosses bare
They have planted thorn trees
For pleasure here and there.

Is any man so daring
As dig them up in spite,
He shall find their sharpest thorns
In his bed at night.

Indeed, Irish fairies are said to meet at the hawthorn trees, or live under them. Cut the lone thorn and it may bleed or scream.

By an English rhyme—

Under a thorn
Our Saviour was born—

the power and ownership of hawthorn is made Christian, and the fame of the Thorn in Somerset is associated with hawthorn as the material of Christ's Crown of Thorns.

Here are some recipes for using both the flowers and the fruits of hawthorn:

Haw Jelly

For this jelly—excellent with cold meat—a Northumberland recipe specifies gathering 3 lb. of ripe, red haws. Wash the fruit well and put in a pan with 3 pints of water which you then simmer for an hour, mashing the fruit down often. Pour into a jelly bag and leave to strain overnight.

Next morning, measure the juice and return to the pan with 1 lb. of sugar and the strained juice of one lemon to each pint of juice. Boil until the jelly is well set when tested.

Another splendid way to eat haw jelly is with cream cheese.

Hawthorn Brandy Liqueur

The strong, almond-scented May blossom makes a fine liqueur.

Cut the hawthorn flowers when in full scent, using scissors, and taking the flower heads only—not the small stems. Pack the flowers into a bottle with a wide mouth, shaking them down loosely—but do not press or bruise the flowers.

Shake a little fine sugar (two tablespoonsful to a pint bottle) over the flowers, fill up with brandy and cork tightly. Put the bottle into full sunshine until warmed through; then store in a warm, dark cupboard.

During the first few weeks shake the bottle gently so that the sugar is dissolved and evenly distributed. After that stage, let it stand unmoved for at least three months. Decant the liqueur gently into a small bottle and cork securely.

Hawthorn Suet Roll

Four or five generations of a cottager's family at Wymeswold, near my present home on the Leicestershire–Nottinghamshire border, have enjoyed an annual 'spring dinner' of hawthorn buds prepared as follows:

Make a fine and light suet crust, season it well with salt and pepper and roll it out thinly and as long in shape as possible. Cover the surface smoothly with the green hawthorn buds (or young leaves will do), patting them down lightly. Then cut a rasher of bacon into very fine strips and lay across the hawthorn buds.

The next stage is to moisten the edges of the crust and roll it up tightly, sealing the edges as you go. Tie in a floured cloth and boil or steam for at least an hour. Unroll on to a hot plate, cut and serve in thick slices with gravy. You will find this 'hawthorn roll' very tasty.

Local Names

Numerous and picturesque names for the hawthorn bush include: Azzy-tree, Buckinghamshire; cheese-and-bread tree, hag-bush and quickthorn, all Yorkshire names; heg-peg bush, Gloucestershire; hipperty-haw tree, Shropshire; Holy Innocents, Wiltshire; quick (especially when used for hedges) and whitethorn are common names in many counties; shiggy and skeeog, Ireland.

The hawthorn *fruit* names include agarve, Sussex; aggle, Devon; bird's eagle, Cheshire; bird's meat, Somerset; cuckoo's beads, Shropshire; hag and haw are common local names in many counties; heathen-berry, Cheshire; hog-berry in southern and western counties; and pixie-pear in Dorset and Somerset.

Mountain Ash (rowan)

This is one of our hardiest trees and, with the birch, is found at higher altitudes in Britain than any other species—sometimes at 900 m (3,000 ft.) above sea level. In Scotland and the North of England the mountain ash is called rowan, and this is one of our most interesting tree names—connecting ancient and superstitious northern customs not only with the old Norsemen but also with the Hindus who spoke the Sanskrit language.

The word rowan is linked with the Old Norse *runa*, a charm, it being said to have the power to ward off the effects of the evil eye. In earlier times *runa* was the Sanskrit word for a magician and *run-stafus* were staves cut from a rowan tree and upon which magic symbols were inscribed.

In Teutonic mythology the mountain ash is associated with Thor, the god of thunder, and it is one of the few trees of Iceland associated with a proverb, in this case: 'The rowan was the salvation of Thor'.

This is derived from a story in early Icelandic literature of Thor visiting the giant Geirroth, in the land of the giants. On his journey, says the strange story, Thor had to cross the Vimur, a blood-filled river dividing the world of men from the world of giants. By catching hold of a rowan branch Thor was able to reach the far bank. According to E. M. Wright, in his *Rustic Speech and Folklore* (1913), there is evidence to suggest that the rowan was Thor's wife; the berries were sacred to Ravdna, wife of the equivalent thunder god of the Lapps.

Rowan Folklore

In Yorkshire, May 2nd used to be called Rowan Tree Day, or Rowan Tree Witch Day, and rowan branches were hung on house doors and windows to ward off evil spirits. Also in Yorkshire and Lancashire, witch-wands or divining rods, were made of rowan; and in Cornwall it was believed that a twig of rowan kept in the pocket prevented elf-induced afflictions of rheumatism.

In South Wales children would dress the Priest's well with rowan and cowslips to keep the witch away from those families who drew water from the well (Francis Jones, *The Holy Wells of Wales*, 1954).

The folklore of rowan is vast, but this short selection would be incomplete without referring to the border ballad *Willie o' Douglas Dale* (*English and Scottish Popular Ballads 1882-98*, by F. J. Child):

Willie has gone off with the King's pregnant daughter to the quiet woods and the girl—wishing to ward off the imminent dangers of childbirth—says,

O had I a bunch o yon red rodding
That grows in yonder wood
But an a drink o water clear
I think it would do me good.

Bait for Birds

The mountain ash (Latin name *Sorbus aucuparia*) is botanically related to the whitebeam and wild service tree, and is so called because its feather-shaped, compound leaves are similar to those of ash—but the edges of each leaf on the mountain ash are toothed. You will often find the tree growing high up in rocky crevices where birds that eat the berries have dropped its seeds.

Mountain ash

Besides being called the rowan in the north, another interesting name is fowler's service tree—this refers to the use of the berries as bait for trapping birds. Everywhere the mountain ash grows—and many are now planted in gardens and along town streets—the orange berries are a favourite with thrushes and blackbirds, who rapidly strip the trees of fruit (Nature's way of making sure the seeds, by passing through the birds' intestines, will be spread far and wide and will germinate quicker than if the whole fruit lay in the soil for some months).

Widespread in woods, on the moors and in rocky places, especially in the north and west of Britain, the mountain ash is a small tree, up to about 9 m (30 ft.), with a smooth, greyish-brown bark. It is easily recognised in winter by its exceptionally large purple-coloured oval buds, with a long point to them (unlike those of the ordinary ash, the buds are set singly on the twigs, not paired). In May come the clusters of creamy-white, fragrant flowers, and by August or September the flower heads have ripened into bunches of berries, brilliant orange outside and yellow inside, the size of holly berries.

If the birds have not found the particular tree, rowan berries can be seen until January—but they are best picked in October when they are fully-coloured but not mushy. They make a lovely sharp-flavoured, dark orange jelly which is extremely good with game and lamb—and is the traditional Scottish accompaniment to grouse.

Bitter Rowan Jelly

A recipe from Wigtownshire: After they have been just nipped by autumn frosts, pick your rowan berries clean from the stalks and stew them down to as near a pulp as you can, with enough water to cover the berries, and a pinch of ginger. Crush and strain the pulp and then boil again for half an hour with two-thirds of its weight in sugar. Put into small jars, and as it cools, the jelly will become firm. Equal parts of crab apples and rowan berries are also used for making this piquant jelly.

31

Rowan Wine

When the berries are fully ripe is the time to make this wine, and the following is a Devon recipe (in which county, incidentally, the mountain ash grows thickly in the hedgerows):

To each quart measure of berries add a quart of boiling water and a small piece of bruised whole ginger. Let them steep for 10 days, stirring well each day. Next, strain and add to each quart of liquid 1 lb. of cubed sugar. When the sugar has dissolved, bottle up—but do not cork tightly until fermentation has ceased.

Ash Pickle

Not the mountain ash, or rowan, but the ordinary ash tree of the hedgerows (Latin name *Fraxinus excelsior*) is very familiar, with its 'ash keys'—the single-winged seeds which hang in bundles from July onwards and from which pickle can be made. John Evelyn recommends boiling the young, green 'keys', changing the water and boiling again—then pickling under hot spiced vinegar.

Local Names

In the north, the mountain ash is known by the Scandinavian names of rowan or roddin; also as quicken, wicken and wiggen. In the south and west, as quickbeam or whitty.

Cornwall and Devon villagers call the mountain ash by the name 'cares'; cock-drunks, Cumberland; dog-berries in the Lake District; poison-berries, Somerset and Yorkshire; chitchot, Wiltshire; shepherd's friend, Dorset; whistle-wood, or witch-hazel, Yorkshire.

Wild Cherries

There are three species of wild cherries in the British Isles. These are the dwarf cherry (*Prunus cerasus*), the gean (*P. avium*) and the bird cherry (*P. padus*).

The gean is the most widely distributed and will therefore be described first. It is a tree which attains an average height of 12 m (40 ft.) or more, and has short, stout branches. It is common in hedgerows and woods, particularly beech woods. Leaves of the gean are large and oval-shaped, and in spring they are of a bronzy-brown colour, afterwards changing to pale green.

The gean's spring beauty is short-lived. Its delicate white blossoms fall within a week, and by July each flower has developed its small, shiny black fruits. These can be either sweet or bitter, but have very little juice and stain the fingers. The fruits look rather like very small, dark reddish-black cultivated cherries. They are thought to be the original wild stock from which our modern Black Hearts and Bigarreau cherries have been grown.

Outside flowering and fruiting times, the gean tree can best be identified by its dull, purplish-grey bark with horizontal bands of pale brown, corky breathing pores. If the wood is damaged it bleeds a light brown resin which seals the wound.

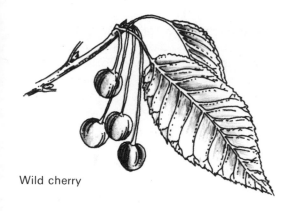

Wild cherry

Wood of the gean is easily worked and can be given a high polish—one of the reasons why it is popular with cabinet and musical instrument makers. It is the fruit of other wild cherries, notably the Morello cherry described below, which are used for wine making and other purposes.

Centuries ago, cherry tree gum, which country children sometimes suck, was recommended for 'coughs, a good complexion, good appetite and keen sight'. Herbalists also believed that, dissolved in wine, gum from the cherry tree helped to dispense stones in the bladder.

Dwarf (or Wild) Cherry

This is a much smaller shrub than the gean. It is also rare and usually only to be found on rich soils. Neither is it found farther north than Yorkshire. The common name of *Prunus cerasus*

is shared with the Morello cherry and the fruit is used commercially for making liqueurs—hence its other name of the brandy cherry.

This cherry is more bush-like than tree-like and sends up a large number of suckers around its main stem. The branches are slender and drooping; its flowers more cup-shaped than those of the gean; its round cherries are red and their juicy flesh is acid to taste— (another identifying feature is that the juice of a Morello cherry does not stain like that of the gean).

The Bird Cherry

This variety makes a tree up to 6 m (20 ft.) high. Its fruits are small, black and very bitter—with a wrinkled stone inside.

Although it has no culinary use the following facts about the tree will help identify it from the useful varieties: It is quite common in woods and hedgerows, mainly in the north of England—and not usually found any farther south than Leicestershire. It can easily be distinguished from the gean and the wild cherry by its long racemes (spikes) of smaller flowers instead of clusters—and its petals in late April or early May look as if their edges had been gnawed.

Cherry Wine

Provided it is kept for at least six months before using, wine made from Morello cherries is a good drink and popular in Kent. (I was given the following recipe when, as a very young man, I worked on the editorial staff of a newspaper at Ashford in that county):

After stalking and washing the fruit, place in a container with 1 pint of cold water to each pound of fruit. Stir three times a day for 10 days, then mash well with your hands and leave for another 10 days, again stirring each day. Tie muslin over another container and suspend a colander over this so that the bulk of the fruit is retained in the colander when the liquid strains through.

Do not squeeze or hurry it, but when all is strained, measure the liquid and to each 2 pints add 1 lb. of granulated sugar. When all the sugar is dissolved, bottle up—and leave to ferment. When fermentation has finished, cork tightly and store.

The method of making **cherry brandy** is as follows: Pack an earthenware jar with dark Morello cherries, sprinkling crushed brown sugar candy. Fill up the cask with brandy and tie down with a bladder. Keep several months before bottling.

Local Names

For the gean: Mazzard, in many parts of the country—particularly the west—also merries; crab-cherry, Buckinghamshire; gaskin, Somerset, Sussex, Kent.

Historical, and very attractive, names for the fruits were mazzards and merries, and these gave their names to the tree.

Bilberry

This is one of the most common of our moorland and mountain plants, growing as a low, spreading shrub among the heather. Whortleberry is an alternative name for the same 450 mm (18 in.) high shrub and this name is a corruption of myrtleberry—from the plant's Latin name *Vaccinium myrtillus*. Greenish-pink, bell-shaped flowers appear on bilberry plants from April to June and are followed in July by round damson-blue berries covered with a fragile grey bloom. These berries, which often stay on the plant until September, are an important part of the diet of grouse and other moorland birds.

Although bilberry is abundant on heaths and moors it is not very common in the south and east of England. Bilberry has hairless twigs and oval, slightly toothed, bright green leaves.

The fruits are juicy and often used for jam—but gathering any quantity requires a lot of patience and the careful searching of a considerable area of land even where the bush is prolific (most commercially-sold bilberries these days are imported from Poland).

On some moors there are bilberry 'rights', and parties of villagers, and often gypsies, go 'up on the mountain' for a day to fill many baskets with the fruit. In Scotland and Ireland bilberries are used for dyeing and in the Hebrides the leaves were formerly dried as a substitute for tea.

If you pick bilberries, select the best by hand and eat these raw with sugar and cream. To wash them, turn the berries into a deep jar and swish the water lightly around—you will find the unwanted bits of leaves and twigs will float up and are easily skimmed off.

Bilberry Tarts

Considered by many connoisseurs to be the best on earth! For tarts (as the very juicy berries 'go down' so much) it is permissible—says Dorothy Hartley in her *Food in England*—to put the white fluff of roast apples into the dish, under the bilberries, to sop up the juice and give the tart a little more substance. In Lancashire, where they make 'fruit between two skins', and also in Derbyshire, a mint leaf is placed here and there on top of the bilberry layer.

Bilberry

Bilberry Pie

As these wild fruits are very succulent, they do not need much extra liquid added when cooking.

Bilberry pie, virtually an American national dish—where bilberries are called blueberries—and which some Yorkshire miners like to take down the pit, is simply berries sprinkled with sugar and lemon juice, baked inside a double crust pie.

Summer Pudding

The well-known cold sweet called summer pudding can be made from any ripe soft fruit (often red or black currants, or blackberries) but many countrymen would agree that the favourite is made with bilberries. Here is a recipe from the late Alison Uttley whose childhood was spent in a Derbyshire village where the springy cushions of plants in Bilberry Wood provided fruit for her favourite sort of summer pudding:

Simmer the bilberries with sugar until soft. Then line a basin closely with strips of bread to make a close-knit mould without cracks inside the bowl. Into this pour the hot fruit and enough of the juice to keep the fruit fairly stiff and to saturate the bread.

Next cover the top with a 'lid' of bread, without crusts. Finally, place a saucer on the top with a heavy weight and leave the pudding all night in a cold place, for the juice to permeate the bread and the whole pudding to congeal to firmness.

The following day it can be carefully turned out and served with thick cream or egg custard. Cheap, nutritious and very tasty!

Local Names

In a number of regions blackberry is a common name for the bilberry—from the old Scandinavian *blaa*, meaning dark blue. Blueberry, in Yorkshire, Cumberland and Ireland; boylocks, Scotland; arts and cowberry, Somerset; hurts and whorts, in many parts of the south and west country; and wimberry (i.e. wine berry) in the west and north Midlands.

Cranberry, Cowberry and Cloudberry

Cranberry

Now largely confined to the north of England, southern Scotland and Wales, this low-growing shrub used to be common everywhere before agricultural drainage of marshy land deprived the plant of its wet habitat.

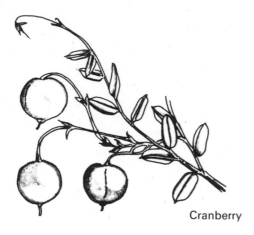

Cranberry

The fruit is mentioned, as fenberry, in a herbal of 1578 and the *Child's Guide* of 1850 says:

Cranberries are grown in the fens, but Longtown in Cumberland grows the most richly flavoured.

The red cranberries, often speckled with brown, ripen in August and September, and are much easier to see than the rest of the plant which has wiry stems and small, rounded evergreen leaves and inconspicuous pink flowers. (Commercially-bought cranberries are larger than the wild ones—Latin name *Vaccinium oxycoccus*—and come from a different American species.)

Cranberries are not eatable raw but they make, among other things, an excellent sauce (indispensable with turkey) which can be kept throughout the winter without 'going off'. Cranberry recipes were preserved by the British settlers in America and only later brought back to us as something new.

Cranberry Chutney

Pick over 2 quarts of cranberries, put them in a pan with $3\frac{1}{2}$ lb. of white sugar, 1 lb. of stoned, coarsely chopped raisins, the thinly pared and finely chopped rinds of two oranges, and $\frac{1}{2}$ lb. of chopped onions.

This Northumberland recipe goes on: Add $\frac{1}{2}$ pint of vinegar and the strained juice of the oranges, plus $\frac{1}{2}$ oz. of mustard seed and a level teaspoonful each of ground ginger, powdered cloves and cinnamon together with a little salt and pepper. Boil until thick, then bottle.

Cranberry Sauce

Stew 1 lb. of cranberries in a teacupful of water until tender. Press through a sieve and return to the pan with $\frac{1}{4}$ lb. of sugar. Reheat and add two glasses of port wine before serving.

Cranberry Jam

An Aberdeenshire method for freshly-boiled jam whenever you fancy it is simply to cover the cranberries with water brought to the boil, then boil the mixture for 10 to 15 minutes. Put in large jars and cover. When you want the preserve you merely boil up again with sugar.

Local Names

Crane, Northumberland; crone, Lancashire; maeberry and mossberry, Yorkshire; moss-mingin, Scotland.

Cowberry

This dwarf shrub (Latin name: *Vaccinium vitis-idaea*) often called mountain cranberry, is, as its Latin name indicates, a relative of the cranberry. It grows on moors in the north country, and its evergreen leaves yield a yellow dye.

Cowberries

Like cranberries, the red fruits are too sharp to eat raw, but they make—with apple added for the pectin—a good jelly. (It is interesting that although the cowberry's botanical name means wine of Mount Ida, a mountain in Turkey, it is a native of northern Europe and is unknown in Turkey.)

———

Another plant of the northern heather moors is the **crowberry**—certainly a confusingly similar name but quite distinct (its Latin name is *Empetrum nigrum*) and it has black, not red, fruits. Poor eating quality, but used in very cold climates as a source of vitamin C.

Cloudberry

Even more dwarf-growing than the crowberry, and again an inhabitant of damp northern moors, is the cloudberry (*Rubus chamaemorus*). It is a relative of the blackberry, and male and female flowers are borne on separate plants.

Cloudberry

Large, orange-coloured berries appear among the kidney-shaped leaves in the autumn, and these can be used for any dish normally made from blackberries or wild raspberries.

Raspberry

Often, raspberries found in the wild are bird-seeded from gardens but the plant (Latin name *Rubus idaeus*) is—like the strawberry—a genuine wild native of the countryside.

The raspberry is a close relative of the blackberry. It has a special flavour, particularly in the wild, and is considered by many epicures to be the best soft fruit of all.

The wild raspberry can be found in all parts of the country growing up to 1.8 m (6 ft.) tall in hedgerows, heaths and some woods. They do, however, grow more vigorously (and the fruits are sweeter) in the hilly parts of the north and west.

Raspberry stems, called canes, are round, downy and have a few small prickles on them—in Scotland the plants are called rasps, presumably with reference to the roughness of the stems, although raspberry prickles are nothing like as many or as strong as those on the blackberry.

Note to gardeners: A few wild raspberry canes brought in and cultivated soon make large fruit, and are very healthy stock.

Wife, into thy garden, and set me a
*   plot*
With strawberry roots, of the best to
*   be got:*
Such growing abroad, among thorns
*   in the wood,*
Well chosen and picked, prove
*   excellent good.*

The barberry, raspberry and
*   gooseberry too,*
Look now to be planted, as other
*   things do:*
The gooseberry, raspberry and roses
*   all three,*
With strawberries under them, trimly
*   agree.*

Thomas Tusser
16th century

The raspberry is usually the first soft fruit to ripen, sometimes as early as the end of June. Indeed, the time to go out and find your wild raspberries is—as Wordsworth has it in his verse *Foresight*—'as soon as spring is fled'.

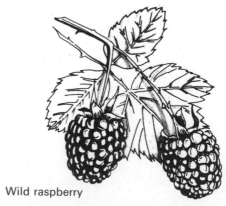

Wild raspberry

Everyone is familiar enough with raspberry jam, but for some of the unusual—and equally delicious—ways of enjoying this fruit we must again search farmhouse and cottage recipes, some very old—some new (like one below for raspberry ice, which needs a refrigerator).

Raspberry Ice

First, sieve 1 lb. of ripe but juicy raspberries. Next cook $\frac{1}{4}$ lb. of sugar and $\frac{1}{4}$ pint of hot water for about five minutes until all the sugar is dissolved —then add the sieved raspberries, a quarter of a tablespoonful of lemon juice and a pinch of salt, and stir well.

After allowing to cool, pour the mixture into an ice cube tray and start freezing. When the mixture is fairly thick but not quite frozen, take it out and whip it very quickly until it becomes really light.

Wash the ice cube tray, line with a piece of plastic or aluminium foil, pour in the mixture and finish freezing it. To serve, whip $\frac{1}{4}$ pint of double cream. Turn out the raspberry ice on to a plate and decorate with the whipped cream.

Raspberry Fool

Warwickshire once more provides me with an excellent recipe.

Stew 1 lb. of raspberries with the juice of 1 lemon and $\frac{1}{4}$ pint of water; when soft, press through a fine sieve. Add $\frac{1}{4}$ pint of custard and 1 oz. of sugar, and stir. Next, whip $\frac{1}{4}$ pint of cream till thick but not stiff and fold into the fruit. Serve in individual glasses.

Medicines

Years ago, many villagers made their own cough mixtures, liniments and ointments much as the earlier herbalists did. Recipes were handed down generation by generation, and raspberry vinegar was one of these home remedies, a delicious one at that. Occasionally one can find a shop that still sells it today.

From Derbyshire comes this recipe:

Put 2 quarts of ripe raspberries into a jar with enough white vinegar to cover them. Stand for 24 hours, then put both fruit and vinegar into a pan and bring to the boil. Next strain, and to each pint of juice allow 1 lb. of sugar. Put it all into a pan and boil for 20 minutes, stirring well.

Bottle when cold and store in a cool, dry place. *Blackberry* vinegar, also useful for a cold, is made in a similar way.

Raspberry Jelly

Wring out a strong linen bag in boiling water and fill with bruised berries. Hang it in a hot oven, or before the fire, until the juice runs and then press strongly. To each 4 lb. of juice add 5 lb. of hot crushed sugar and boil for five minutes. Pour into glazed earthenware pots so that the jelly, which sets when cold, can be easily turned out.

Cut into slices, the jelly was a favourite in the 17th century for garnishing dishes.

Raspberry Footnote

If you can find no more than a handful of wild raspberries, use them for stuffing game birds or to make a cold 'summer pudding' (see recipe under Bilberry).

Strawberry

The wild strawberry, bearing dainty white flowers from April until July, grows in waste land, open woods and hedges throughout the British Isles, especially on lime-rich soils. Its fruits are smaller, but sweeter than garden varieties.

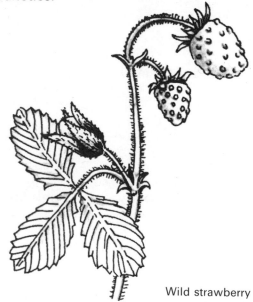

Wild strawberry

Another species, called barren strawberry (*Potentilla sterilis*) does not produce fruits worth eating. It also has white flowers, sometimes as early as February, but it can be distinguished from the useful wild strawberry (*Fragaria vesca*) by its leaves, which have a blue tinge and spreading hairs beneath. The barren strawberry is found everywhere except in Scotland.

You will need to look quite carefully for wild strawberries in rough grass where the small, drooping red berries are often hidden by the toothed, shiny green leaves grouped in threes. But wild strawberries are well worth the picking—and best eaten fresh, with cream. A pleasant jam, or a filling for tarts, can also be made with wild strawberries, and they need only a little sugar.

The fruit is historically interesting. The strawberry was the fruit of Venus and the Virgin Mary. Strawberries are featured in the centre panel of the medieval painting by Hieronymus Bosch called 'The Garden of Earthly Delights', on view in the Museo del Prado, the state gallery in Madrid. This strange painting shows naked men and naked girls engaged in the art of weaving whilst, at the same time, enjoying tender sexually symbolic actions—a nude man offers a nude girl a huge ripe strawberry double the size of his own head—an allegory in paradisical eroticism, painted for the sect of Adamites which sought to re-establish the innocent state of man at the time of the creation.

Before 1600, when American wood strawberries (*Fragaria virginiana*) were introduced to Europe as the forerunners of the big, cultivated strawberry of today, they were simply wild plants brought in from the woods and grassy banks, and cultivated.

Shakespeare has a reference to strawberries in *King Richard III* when Gloucester speaks to the Bishop of Ely (obviously wishing the Bishop would depart!) saying:

When I was last in Holborn,
I saw good strawberries in your
    garden there;
I do beseech you send for some of
    them.

These strawberries would, at the time, have been the wild variety transplanted from woods in spring. The bishop's garden was in the vicinity of what is now called Ely Place, Holborn, land which was a detached part of the see of the Bishop of Ely, the city in the Cambridgeshire fens. (John Gerard, 1545–1612, herbalist and writer on gardening also had a large physic garden at Holborn where he practised as a barber-surgeon.)

A Classic Dish

Half fill a deep, cold bowl (an old punch-bowl is excellent) with cream and whip it slightly—but do not make the cream too stiff. Drop in as many strawberries as the bowl will hold, stir as you go, mashing slightly and then leave to stand for an hour. It will then be a cold, pale-pink cream. Crust it over with sugar and 'serve forth [advises Dorothy Hartley] in June, on a green lawn, under shady trees by the river'. The cream is perhaps best served with fruit fillings in sponge cakes.

We may not all be able to do it so elegantly, but without doubt this is *real* strawberries and cream. Its praises have also been sung by Andrew Boorde, a 16th-century Carthusian monk who, after becoming Bishop of Chichester in 1521 was, eight years later, freed from his vows to become a physician and writer (incidentally it was Boorde who, after travelling through nearly every European country wrote *Fyrst Boke of the Introduction of Knowledge*—the very first Continental travel guide).

This is Andrew Boorde's comment on red strawberries and cream:

Rawe crayme undecocted, eaten with
    strawberys or hurtes is a rurall
    manner banket.
I have knowne such bankettes hath
    put men in jeopardy of they lyves.

Bavarian Pudding

Try to have enough patience to pick 1 lb. of wild strawberries and, on a warm, early summer day, give the family a really unusual pudding treat. Its full name is sweet Bavarian rice pudding, and this is how to make it:

Put 1½ pints of milk and a few thin strips of lemon rind into a double saucepan. When hot, stir in half a cupful of rice and a pinch of salt, and cook until the rice is tender—the milk should be very nearly absorbed, leaving the rice moist.

Next, add to the hot cooked rice one teaspoonful of vanilla essence, half a cup of sugar, and 1 oz. of gelatine soaked in cold water—and mix very carefully.

When the mixture is beginning to set, fold in ¼ pint of cream whipped stiff. Pour into a ring mould and place in the refrigerator. This is served with the strawberries, sweetened and sliced, arranged tastefully over the pudding— a mouth-watering sight indeed!

Red and Black Currants

Both the red currant (Latin name— *Ribes sylvestre*) and the black currant (*Ribes nigrum*) are native British shrubs, although some plants found in the wild are seed-sown from garden varieties. Nevertheless, the wild red currant is often found at the sides of rivers and streams, especially in uncultivated fenland. The black currant is more uncommon in its true wild state—it can

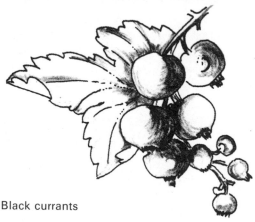

Black currants

be distinguished from its related red currant by its large, scented leaves, a few of which when dried can give a pleasing special flavour to a pot of tea. (The leaves of the red currant, which give a yellow dye, do *not* smell when crushed.)

The red currant is a bush growing to about 1.2 m (4 ft.) high and is common, but localised, in woods and hedgerows near water. Its toothed leaves are broken into three or five lobes, and the flowers are small, light green in colour and drooping. Fruits of the red currant, ready from July, are round and shiny red.

Beware Confusion

Here a warning must be given, especially to children. The guelder rose (*Viburnum opulus*) is also a waterside shrub with lobed leaves and bright red fruit from July onwards; it is common in hedges and woods throughout Britain but **the guelder rose berries will produce sickness if eaten raw** (cooked they can be made into jellies, and in America are used as a substitute for cranberries in sauces). However, these berries lack a 'tail' as on those of the red currant, and look heavy and waxy beside the translucent skins of the currant.

Red currants

Red Currant Jelly

Red currants contain a lot of pips, and for this reason jelly is the most popular way of using them.

Pick over 6 lb. of fruit, leaving it on the stalks but removing all bits of leaf. Wash the currants well in a colander with plenty of cold water and, after draining, put the fruit with 2 pints of water into the preserving pan and heat through slowly—keep simmering until the fruit is cooked and really soft.

Mash the currants with the back of a wooden spoon, then strain through a jelly bag (a pair of old tights is, as I said before, a cheap substitute!) and then leave to drain for several hours. The following day, measure the juice, bring to the boil, add 1 lb of sugar for each pint of juice and stir until dissolved. Boil up again and continue to boil for 10 minutes. Test, and if ready, pot up into small, hot jars and cover down at once.

A Sauce for Puddings

One tablespoonful of red currant jelly and raspberry jam, plus $\frac{1}{4}$ pint of water, should be boiled together for five minutes. Mix a teaspoonful of corn-flour in a little water, stir it into the sauce and boil for a minute or two—then strain the sauce over any kind of plain steamed pudding.

Red Currant Wine

To each 4 lb. of fruit (garden if you can't find enough wild) add 5 pints of boiling water. Crush the fruit, cover and allow to stand for four to six days. Strain off the juice and to each gallon stir in 3 lb. of sugar and $\frac{1}{4}$ oz. of yeast, liquefied with a little of the juice. Pour into jars to ferment, filling to the brim. Cover with greaseproof paper with a slit made in it—this is to keep small the area exposed to the air.

Keep the jars topped up to the brim, and allow to ferment in a warm room for four to six weeks until the bubbling stops. At this point, stir thoroughly; leave for three days and then pour through filter paper in a funnel into wooden casks or bottles and cork loosely—until fermentation has ceased, when you can cork tightly.

Red currant is a delicious wine if kept for nine months or more in a cool dark place before decanting.

Black Currant Wine

An old Buckinghamshire recipe starts with picking 4 gallons of black currants and putting these in a large earthenware jar (a plastic container is more likely today) with a cover to it. Boil $2\frac{1}{2}$ gallons of water with 6 lb. of cubed sugar. Carefully remove the scum as it rises on the liquid and pour, still boiling, on the currants.

Let the lot stand for two days, then strain into another vessel, return to the jar (or plastic bin) and leave for two weeks to settle before bottling off. Excellent for winter colds and coughs.

Black currants can also be dried and in this form they are one of the ingredients of pemmican—meat cakes formerly eaten by North American Indians and later by Polar explorers.

Local Names

(For the red currant): wineberry, Yorkshire, Cumberland and Scotland; gazel, Surrey and Kent; squinancy-berry, Essex, Lancashire and Cumberland; wineberry, Scotland.

The word *gazel* is from the French groseille, a red currant.

Gooseberry

This is familiar enough as a garden soft fruit—but the gooseberry also grows wild in woods and hedgerows. They are, however, very localised in the north and rare in Ireland.

The name gooseberry (Latin—*Ribes Uva-crispa*) means what it says—a berry eaten as a green sauce with geese (the sauce was made with the addition of sorrel and sugar).

It is a many-branched shrub, rarely above 0.9 m (3 ft.) tall, with lobed leaves and drooping green flowers in March. The egg-shaped and hairy fruit appears from July onwards, and is greenish-yellow.

Gooseberry

Gooseberry Sauce

Besides the use of gooseberries in a green sauce to go with young geese, they make a splendid sauce to eat with mackerel, or as an adjunct to cold meat.

Here is a recipe I found in a North Wales inn:

Put 2 quarts by measure of gooseberries into a lined pan together with 2 cups of vinegar, 3 lb. of brown sugar and a dessertspoonful each of cinnamon, cloves and allspice. Cook slowly for two hours, being careful that it does not burn. Put into wide-mouthed bottles and seal down tightly.

The green sauce of olden times was made by pulping the leaves of sorrel (sorrel—*Rumex acetosa*—is an acid plant of the meadow once used as we use lemons) and mixing with sugar and vinegar. The boiled leaves of sorrel were also eaten, or added to ale.

Gooseberry Pickle

To make gooseberry pickle the same Welsh document previously quoted gave the following method:

Boil in an enamel pan 3 lb. of gooseberries, $1\frac{3}{4}$ lb. of sugar, $\frac{1}{4}$ oz. ground cloves and $\frac{1}{2}$ pint vinegar. Whilst cooking, stir the mixture until it is the consistency of jam. Put into jam pots and cover when cold.

Gooseberry Wine

Wash, top and tail 8 lb. of gooseberries, bruise them well (a rolling-pin is a good tool, says an Essex recipe) and place in a container with 2 gallons of water. Mash well and leave covered with a cloth for two days before straining and measuring the liquid.

Put the liquid together with 3 lb. of

sugar for each pint of juice back in the container and stir often until all the sugar has dissolved. Pour into a cask and leave in a warm place until fermentation has stopped—probably about three weeks. Then drive in the bung securely, fit a peg in the vent hole and every day pull the peg out to allow any gas to escape.

When all seems quite still, close up tightly and leave in a cool place for eight months before bottling.

Gooseberry Tansy (about 1700)

Cook a quart of gooseberries with some butter in a covered jar till quite soft. Beat 4 eggs and fill them with a double handful of fine white breadcrumbs and a cupful of sugar—blend this into a gooseberry pulp over a slow heat, stirring gently till the mass is cooked firm (on no account let it get too hot or the custard will curdle).

Turn out on to a hot dish, sprinkle with crushed sugar, and serve with hot cider or melted apple jelly.

It is a little tricky to make a really good tansy, the knack being to use only just as much breadcrumb and egg as will take up the buttery liquid from the cooked fruit, and make the same 'bind'. It may be found necessary to add another piece of butter if the fruit is very dry, or more crumbs if the fruit is soft. The completed dish should be the consistency of a solid omelette.

Gooseberry Raised Pie

Nottinghamshire folk used to make a pie, raised like a pork-pie crust, but filled with green gooseberries. After baking, melted apple jelly was poured in through a slit in the lid. The pie was always served cold, as it cut like a pork pie, solid, with firm-set jelly between the green gooseberry centre and the white firm crust.

Gooseberry Fool

This old English sweet was made with whipped cream, but nowadays we compromise on custard. Put the green gooseberries into a jar and bake (covered closely) till soft pulp; while still hot, stir in a pat of butter and enough sugar to sweeten. Let this cool (you may rub through a sieve, but this is not necessary unless the skins are tough). When cold, fold into an equal quantity of cold whipped cream, or cold stiff custard. It is important that the gooseberry pulp and the cream be exactly the same consistency so that they combine as a soft green cloud.

Medlar

A curious, long-lived tree with a curious fruit—medlars need to be half rotten, or 'bletted' like an over-ripe banana, before they are edible.

The medlar tree (Latin name—*Mespilus germanica*) has a crooked trunk, with gnarled grey bark. It is probably a native of Britain and is often found in old cottage orchards or wild in the hedgerows of the south and west. Its twisted shape is caused by the wood's sensitivity to the wind. John Gerard knew the medlar in the sixteenth century as a tree found 'oftentimes in hedges among briars and brambles'.

The medlar

In its wild condition the medlar is a much-branched and spiny tree, up to 6 m (20 ft.) high and looking like a hawthorn. Its large, pale green oblong leaves are almost stalkless and have downy surfaces. The solitary white flowers, which have five frilled petals, open in May. The flower develops into a pear shaped, yellowish-brown fruit about an inch across, with the calyx (the leaf-like 'envelope' of a flower) protruding from the head of the fruit.

In the Mediterranean, medlar fruits can be eaten straight off the tree but in our colder climate they stay hard and sour until mid-winter. In October they can be picked and stored (in moist bran is a Cheshire custom) until they become soft, when the brown flesh can be scraped out and eaten with sugar and cream; or medlars can be made into jelly—or even baked on a shallow dish like apples, with butter and cloves.

In Aelfric's *Glossary* of about AD 1000 the medlar is given its old English name (from an obvious likeness!) of 'mespilus openaers', and Shakespeare writes in bawdy fashion about the medlar in *Romeo and Juliet*. Thus Mercutio, in Act II:

If love be blind, love cannot hit the mark.
Now will he sit under a medlar tree,
And wish his mistress were that kind of fruit
As maids call medlars when they laugh alone.
O Romeo, that she were, O that she were
An open et cetera, thou a poperin pear!

Medlar Jelly

To every 4 lb. of fruit, washed and cut up, add 3 pints of water and 6 table-spoonsful of lemon juice (or two level teaspoonsful of citric or tartaric acid). Crush and simmer for at least an hour (says an Essex recipe). Strain as hot as possible and very finely and add $\frac{3}{4}$ lb. of sugar to each pint of juice.

Bring to the boil quickly after adding the sugar gradually. Test for setting after five minutes, remove scum and pot whilst hot. Seal tightly.

Medlar Comfits

These sweetmeats are most enjoyable with nuts and port, or to eat with coffee after dinner. This is how to make the comfits:

Simmer 4 lb. of ripe medlars in 1 pint of water until soft enough to press through a sieve. Weigh the pulp and add an equal weight of sugar. Cook slowly until the sugar melts, stirring continuously, then let the mixture boil until it thickens into a paste—and stir it all the time with a wooden spoon.

When the paste is like dough, and no longer sticks to the sides of the pan, put it on a pastry-board which has been thickly sprinkled with castor sugar. Smooth the mixture with a knife and leave to dry. Finally, sprinkle with castor sugar, cut into small squares and store in a tin. (The same recipe could be used with dried apricots.)

Quince

The quince is a Mediterranean tree which has been cultivated in most of Europe since Roman times and is occasionally seen growing wild along hedgerows in Britain.

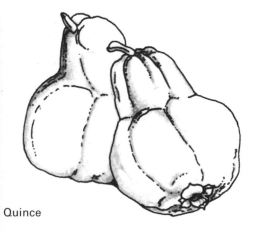

Quince

An apple-like tree, the quince (Latin name—*Cydonia oblonga*) has grey bark and oval leaves, dull green above and greyish and hairy on the underside. Large-petalled pink or white flowers open in April, followed in September and October by large yellow fruits (they go red when cooked) of irregular shape and woolly skin.

Quince Marmalade

An 18th-century Derbyshire recipe:

With each pound of fruit use an equal weight of sugar, a pint of water and a tot of brandy. Wipe, peel and cut up the quinces and leave them in cold water to keep the colour.

Put the rinds in cold water and boil until tender. Strain off the liquid and when cold put the peeled quince into it, weighing a pound of quince to a pound of sugar to a pint of liquid. Boil all until tender, keeping closely covered. Beat with a wooden spoon until you have the right marmalade thickness. Pour into warm pots and 'cover when cold with brandied paper'.

A more modern recipe suggests adding chopped walnuts, stoned raisins and lemon juice after putting in the sugar.

Quince Roast

Originally an even older recipe than the one above for marmalade is the following, dating back to 1700:

'Quinces roasted over hot ashes— scraped out of their skins and mixed with cream and a little sugar and ginger make a dish for children.'

Quince Jelly

Pare and slice some quinces and put in a preserving pan with enough water to float them. Boil until the fruit is reduced to a pulp, then strain through a jelly bag and to each pint allow 1 lb. of sugar. Boil the juice and sugar together for about $\frac{3}{4}$ hour, removing all scum as it rises. When the jelly appears firm, pot up at once into small jars.

A Kentish custom is to make jam with the pulp left over. Put the pulp through a sieve or mash very finely with a wooden spoon. Add $\frac{1}{2}$ lb. of granulated sugar to each 1 lb. of pulp and boil until it sets—keeping well stirred to prevent burning.

Quince and Apple Pie

A book published in 1846 called *The Modern Cook* (it was written by Charles Elme Francatelli) gives this epicurean advice:

'Line the bottom of the pie dish with peeled, cored and sliced apples and add the quince [previously sliced very thin and stewed over a slow fire with a little water, sugar and piece of butter]. Add sufficient sugar to your pie to sweeten it to your liking, stew zest of lemon over the top [this is the rind rubbed on lump sugar then scraped off]. Cover the pie with puff pastry, press it with the back of a knife to scallop it round the edges. Egg the pie over and ornament the top by drawing some fanciful designs with the point of a knife and bake it of a light colour. When done, shake some sugar upon it and use the red hot salamander to glaze it.'

Salamander, as a circular iron instrument for browning cooked foods, is a word not often seen now—today the word usually refers to the small, tailed amphibian like a newt.

Some old cooks advised that, after covering the top of an apple pie with pastry and trimming the edges one should 'pass a giggling iron round the pie'. Just what, one wonders today, was a giggling iron? A lovely term, whatever it meant, but now forgotten.

Barberry and Oregon Grape

Barberry

This spiny shrub, once common in hedgerows even if localised, is now scarce. The reason is that barberry (botanical name—*Berberis vulgaris*) was found to harbour the wheat disease, black rust fungus, and farmers now destroy the plant.

If you should find a barberry bush you will certainly need gloves to get the brilliant red berries as the thorns are very plentiful and sharp. Barberries were very popular with 18th-century cooks, and the shrub has long been cultivated in gardens—for the supposed power of its roots to cure jaundice, as well as for its ornamental value. Its berries can be made into jam or a pleasantly tart jelly.

Over the Mutton

Another way of using the oblong-shaped barberries, particularly if you have only a few, is to place them over roast mutton (particularly if the meat is very fat) for its last few minutes in the oven, so that the berries burst and the juice runs over the meat.

Barberry grows up to 3 m (10 ft.) high, with sharply toothed oval leaves sprouting in tufts along the stems and clusters of small yellow flowers. These flowers have abundant nectar which attracts the bees, and when a bee enters a flower the stamens close over it—covering the bee with pollen which it carries to the next flower it visits. The wood of the barberry is yellow and its inner bark yields a bright yellow dye.

The berries appear in July. They are too sour to eat raw, but apart from the uses already mentioned the berries can also be pickled, or preserved in sugar for use in curries.

Barberry

Years ago barberries were candied for decoration on sweet dishes. Sugar was boiled to syrup point, and the bunches of barberries dipped into the sugar and left for a few hours. They were then lifted out, the sugar reboiled to candy point and the berries put back and left to candy. Barberries were also used to flavour punch, or as a 'fever drink'.

Local Names

Jaundice tree, Cornwall and Somerset; pipperidge is a name common in many counties, including Lincolnshire; wood-sour, Oxfordshire; berber, Scotland.

Oregon Grape

Now more plentiful than the barberry, this evergreen shrub is nevertheless a member of the same family (Latin name—*Mahonia aquifolium*) and is widespread in the south but not common elsewhere.

This shrub, probably naturalised from gardens, grows up to 1.2 m (4 ft.) high, and has bunches of white-bloomed berries which look like grapes. They can be eaten raw or made into jelly.

Whitebeam and French Hales

Whitebeam

This tree, locally common in the south and often used in hybrid form as a roadside tree, is a member of the rose family and is related to the mountain ash (botanical name—*Sorbus aria*).

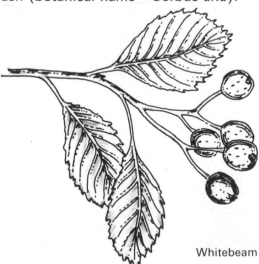

Whitebeam

Whitebeam grows up to 15 m (50 ft.) but does not live for more than about 80 years. The bark is green-grey and the oval leaves are set singly on the twigs. When the wind turns up the whitebeam leaves they appear silver-coloured. Clusters of white flowers in May are followed in October by yellow-brown berries—too sour to be eaten raw. They can, however, be *made into a jelly*.

It is also called the whiteleaf tree because of the whiteness of the undersides of the young leaves, and it was one of these trees that gave its name to a field near Caterham on the Surrey Downs. A rich Victorian built a mansion there and called it 'Whyteleafe' —and today this is the name of the village itself.

French Hales

The Latin name of this close relative of the whitebeam is *Sorbus devoniensis* and as this botanical name implies you are only likely to find this tree in old woods in Devon and a few other localities in the West Country—the Avon Gorge, for instance in Somerset.

In Devon the brown fruit has long been sold and eaten. The word hales is apparently *halse*, i.e. hazel (although this shrub is a different species altogether).

Wild Service Tree

Still in the same *Sorbus* family as the whitebeam and French hales is the wild service tree (*Sorbus torminalis*), but it is unlikely to be found except on rough slopes beside streams in southern England and Wales. It is a small tree with bark broken into squares, similar to hawthorn, hence the old Kentish name of chequers tree since it resembles a draughts board.

The service tree grows only up to 12 m (40 ft.) high, with lobed leaves like those of a sycamore, and its white flowers in May look like those of the mountain ash—but its berries are very different. These come in October, are apple-shaped and greenish-brown in colour. Again like the mountain ash, birds devour these berries.

In the 19th century, wild service tree berries, by the name of chequers, were sold in the markets of southern England. The fruits are small and eatable only when half rotten. The *Flora Vectensis* (Bromfield, 1856) says the chequers fruit was tied up in bunches and sold to children in the markets of Sussex and the Isle of Wight.

Wild service tree

In medieval times the fruits of the wild service tree were used in medicines —Evelyn recommended them for gripe; and water distilled from the flower stalks and leaves for 'consumption, green sickness in virgins and earache'.

Pear

The wild pear (*Pyrus communis*), a member of the rose family, is occasionally found in the edges of woodland in the south of England.

It is the ancestor of many of our cultivated dessert and cooking pears. John Lawrence, a Northamptonshire rector, was a pioneer of garden pear cultivation, and in 1714 wrote an account of his horticultural work in a book called *The Clergyman's Recreation*.

Wild pear

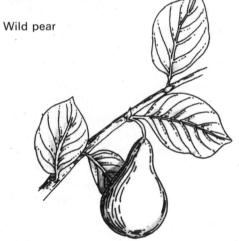

The wild pear, a handsome tree, can be found all over Europe, and is recognised by its upright growth, grey bark divided into squares and brown twigs, some carrying long sharp spines (which are not found on garden pears). The tree flowers in early April, just before the leaves open—a beautiful sight. By October the tree has its greenish-brown pears, hard and sour at first but softening in time and becoming sweeter.

Stewed Pears

From *The Queen's Royal Cookery* (T. Hall, 1707):

'Pare them [but leave whole with stalks on] and put in a pipkin [a small earthenware jar] with so much red or claret wine and water, of each as much as will near reach the top of the pears. Stew or boil them gently till they grow tender, which may be in two hours [in a moderate oven, at say 350°F]. After a while put in some sticks of cinnamon bruised, and a few cloves when they [the pears] are almost done. Put in sugar enough to season them well and their syrup which you pour out upon them in a deep plate.'

Pears and Vine Leaves

A 16th-century recipe states:

'Lay a layer of vine leaves in an earthern pot, and over them peeled and halved pears, and then leaves and then pears again, till the pot be full, and putting some fine sliced gynger, and a few cloves between each layer. (Put not too many cloves, one to each dozen pears unless they be very large.) Fill up the pot with as much [cider] as the pot will hold, and lay some dish or board upon the pears that they do not swim, and cover the pot closely, and stew it all night when the fire be low—or in the oven after the bread be drawn.'

Let the pears grow cold in the pot; they will be soft, golden brown, and should be served cold with whipped cream, or custard and sugar.

Pear Preserve

This modern Gloucestershire recipe is for any kind of pear, and the preserve is very enjoyable eaten with rice pudding or blancmange:

Peel, core and cut up 1 lb. of pears and, with $\frac{3}{4}$ lb. of sugar, boil until cooked—about $1\frac{1}{2}$ hours usually. Depending on whether you make more or less than this quantity, add a teacupful of vinegar to 6 lb. of preserve and boil for 20 minutes. Add crushed root ginger to taste.

Juniper

This is a graceful evergreen shrub of the pine family, but apart from still being widespread on the moors and in the pine woods of Scotland, and reasonably prevalent on the chalk downs in southern England (particularly Wiltshire) it is now becoming uncommon. The reason is that the decrease in the number of rabbits—due to the disease of myxomatosis—means that they are failing to check the growth of the hardier hawthorns and elders where juniper formerly flourished.

The shape of junipers varies with location and climate. In the south they can become little trees; in lowland Scotland, small shrubs, and at greater altitudes in the Highlands the juniper is a prostrate plant.

Juniper has a fibrous red bark, which flakes off like that of the yew tree. The leaves are rigid and end in points, and the leaves are arranged round the branches in rings of three. Male and female flowers appear in the spring and are borne on separate trees—as often is the case with holly—the male flowers small and yellow, the female tiny, bud-like ones.

The berries begin to ripen by late autumn, at first green then—a year later—blue-black and covered with a bloom. They have a pungent flavour and, picked in their second year when black,

Juniper

are used in making gin—which indeed owes its name to this fact; the word gin being a contraction of *genevrier*, the French form of juniper. Juniper berries have long been considered a kidney stimulant—and many gin drinkers convince themselves of its medicinal properties!

Juniper berries are sometimes used in spice pickles.

Latin name: *Juniperus communis*.

(The Virginian Juniper, or Red Cedar as it is sometimes called in America, is botanically called *Juniperus virginiana* and is a much larger plant, often seen in parks and gardens.)

Hottentote Fig

This garden plant from South Africa (*Carpobrotus edulis*) better known by its old name of mesembryanthemum or Livingstone Daisy, is now naturalised in the warm climate of Devon, Cornwall and the Channel Islands. Its matted, succulent foliage and silky pink flowers will be found on the cliffs and rocky banks near the sea.

The magenta, or yellow, flowers are 2 in. across and appear between May and July. Later, the fruits, called figs, are safe to eat (*carpobrotus* is from the Greek, meaning edible fruit) and have a pleasant though acid taste.

The earlier tongue-twister name of mesembryanthemum became, in everyday Cornish speech; Sally-my-handsome.

NUTS

The definition of a nut is a large, hard seed, usually the product of a single flower. Nuts are found only on trees and shrubs because larger food reserves than those of smaller plants are needed to create them.

All nuts are rich in protein and fats. The little hazel nut, for instance, weight for weight contains half as much protein again as an egg. A pound of walnuts can provide 3,000 calories — more than the average adult needs from a day's meals. And for years vegetarians have substituted nuts for other sources of protein needs in human diet.

Always keep any wild nuts you pick very dry, otherwise damp will rot them. Above all, remember that the lives of many of our birds and wild animals depend on nuts—so, please, do not pick them greedily.

 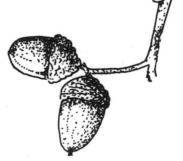

Hazel

Except in stagnant, wet areas, or where the soil is very poor, you will find the hazel everywhere in Britain—and it is perhaps best known for the yellow male catkins carried near the tips of the twigs in February and called 'lambs' tails'. The flowers, which appear before the leaves of the hazel, shed copious pollen and at the same time the hazel bush puts out tiny female catkins, looking like large leaf buds and ending in crimson tassels.

In the autumn these female flowers ripen into groups of two or three round brown nuts, each set between small leaves. Hazel nuts are also known as cob nuts and cultivated varieties producing larger nuts are grown in orchards, particularly in Kent—hence the name of Kentish cobs seen in the greengrocers' shops. (Filberts, also sold in the shops, have long, thin shells as distinct from the hard, round shell of the hazel. The Barcelona nut, imported in winter, is a variety of hazel.)

Hazel (Latin name—*Corylus avellana*) grows in woods, hedges and on wastelands, a spreading bush not often any more than a few feet high, with smooth, brown bark. Neither is the hazel always successful in bearing fruit, and the bushes have to be at least seven years old before they can do so. Sometimes very strong winds, or hard frosts, destroy both the catkins and the crimson stigmas of the female flowers. Modern farm hedge trimming also takes its toll, but hazel left alone will grow into a tree and live for centuries.

Tough but pliant staves of hazel are valuable for making hoops for casks, walking sticks—and water divining rods (which were introduced to Britain by German miners in the 16th century). Hazel is also considered a magic tree, and in Ireland it 'protects against spirits and evil, and abduction by fairies'.

Hazel nuts

Many an Irishman carries a nut in his pocket 'to prevent rheumatism or lumbago', whilst in Devonshire—where it is called a loady nut—folk believe it cures toothache. In Scotland, the double hazel nut, or St. John's Nut, was thrown at witches.

When to Gather

Late September, when the nuts are just turning brown, is about the best time to gather them—and you need keen eyes to distinguish nuts from the crinkly, yellow-tinted autumn leaves. Even at this time the nuts are not fully ripe but it is a case of having to get some of the nuts before the squirrels and jays take the lot! You may be tempted to pick in August when the nuts are abundant but they will be soft and tasteless. When you do pick, go gently since, if hazel nuts are ripe, they will fall from their husks if the tree is shaken.

Keep hazel nuts in their shells and in a cool, dry place. As well as being eaten raw, hazel nuts make a good substitute for almonds, or they can be grated in salads.

A Protein Drink

Like the walnut (see page 61) hazel nuts are rich in protein—they contain 50 per cent more protein, seven times more fat and five times more carbohydrate than a hen's egg.

To make the drink, take two parts of shelled hazel nuts to six parts of milk and two of honey. Grind the nuts very finely and blend in the honey and milk. Serve chilled.

Hazel Nut Meringue

Mix 3 oz. of chopped hazel nuts, 6 oz. of castor and icing sugar mixed plus a pinch of salt and fold into the whites of three beaten eggs. Place gently into a well-oiled or buttered baking tin and cook for two to three hours at 225–230°F until crisp but still white in colour.

From the same Derbyshire source as this recipe comes a reminder of an old farmhouse custom of serving the home-made cream cheese on a doily of hazel leaves 'to add to the beauty of the cheese'.

Local Names

Besides cobnut and filbert already mentioned, the hazel nut is called filbeard in Gloucestershire, Oxfordshire and Northamptonshire; hasketty, Dorset; and woodnut in Yorkshire.

Sweet Chestnut

Like the walnut, this superb tree originally came from Asia Minor and Greece, was introduced to Italy in remote times and brought to Britain nearly 2,000 years ago by the Romans —a staple ration of the Roman legions was a nutritious flour called *pollenta*, made from sweet chestnuts.

The sweet chestnut (botanical name —*Castanea sativa*) is a relative of the oak and now grows wild in England; most common in Nottinghamshire, Norfolk and Gloucestershire, but rare in Scotland and Ireland. It occasionally grows as high as 36 m (120 ft.) and has a 500-year life span—even longer, that is, than the average well-preserved oak. (Stop to think for a moment of the wonder of, among other features of a very big tree, the hydraulic feats performed by trees; no ordinary suction pump can raise a column of water in pipes for more than 9 m (30 ft.) —but trees manage to bring water a hundred feet or more from their roots to their crowns!)

The sweet chestnut has stout, round buds which develop into large, broad and oval leaves edged with saw-tooth points. Its bark is grey and becomes deeply ridged and brown with age. Chestnut catkins open in July, and by October the female flowers develop into triangular, bright brown nuts set in pale green husks covered with soft spines—opening them needs gloves.

Sweet chestnuts are an important item of food in southern Europe, and Evelyn has this in mind when he recommended the nut as 'a lusty and masculine food for rustics at all times, and of better nourishment for husband-men than cold and rusty bacon.'

Sweet chestnut

Eating them Raw

When the ripe nuts begin to fall in October they can be enjoyable eaten raw if the shell and slightly bitter inner skin are removed—but are, I think, best roasted or boiled until soft. Just slit the skins and either put the nuts in the hot ash of an open fire or on the grate close to the fire itself.

A useful old hint is to place with them one nut *uncut*, since when this bursts with a bang you will know the others are ready to eat.

It was a Victorian fortune-telling trick to name the row of chestnuts set along the bar of the fire-grate and the first name to 'pop' was the first lover to 'pop the question'. If the named chestnut jumped into a girl's lap she had him; if he popped into the fire—well, she didn't!

Chestnut roasters, metal containers with a lid and long handle, can sometimes be found in antique shops and markets, and are sensible fireside cookers for winter evenings.

Chestnut roaster

Raw chestnuts are rich and nourishing but will not keep for more than two months—but they can be shelled and dried and later reconstituted by soaking in water.

Chestnut Soup

Cut the ends of 1 lb. of chestnuts, boil for half an hour and drain (cutting saves time in both boiling and peeling them). Remove the outer and inner skins and place in a saucepan with $1\frac{1}{2}$ pints of vegetable stock, one onion and half a stick of celery, and simmer until tender. Pass through a sieve, thicken with 1 oz. of flour blended in $\frac{1}{2}$ pint of milk and cook for another 10 minutes. Season with pepper and salt and serve hot.

Chestnut Cake

Cut, with a sharp knife, the skins of 1 lb. of chestnuts and boil in water for half an hour. Drain, and skin while still hot—then put through a sieve.

Melt $\frac{1}{4}$ oz. of gelatine in a little hot water and strain into a cupful of milk or cream. Stir the mixture into the chestnuts with $\frac{1}{2}$ oz. of sugar. Mix very thoroughly and pack into a plain mould to set. Turn on to a dish, garnish with angelica and pour a little caramel sauce around the cake.

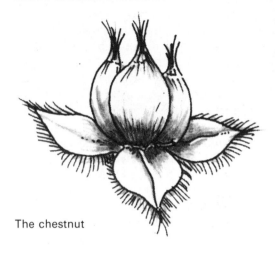

The chestnut

Turkey Stuffing

Slit the skins of $\frac{1}{2}$ lb. of chestnuts and, as with the cake above, boil for half an hour and shell while hot. Press through a sieve and then mix with 2 oz. of breadcrumbs, 1 oz. of grated suet, a tablespoonful each of lemon juice and chopped parsley, a little salt and one egg for binding.

Chestnut Jam

Here is a Cambridgeshire recipe:

Boil 2 lb. of sweet chestnuts until tender, peel and skin them. Then crush through a wire sieve.

Make a syrup with $\frac{1}{2}$ pint of water, sugar and three tablespoonsful of vanilla essence. Next put in the crushed chestnuts and cook gently until fairly stiff. Put into hot glass jars and always store this jam in a dry place.

Orange Salad

Chestnuts and oranges make an attractive fruit salad, pleasant to eat with cake or biscuits.

Boil 1 lb. of chestnuts, draw aside and one at a time remove both skins. Put ¼ lb. of sugar and 1¼ pints of water into a pan plus a few drops of vanilla essence. Boil for five minutes, add chestnuts, and simmer gently until the nuts are semi-transparent and the syrup fairly thick.

Cut the rind and pith from three large oranges and cut out the flesh from the sections. Draw the chestnuts aside when done, cool and add the oranges. Chill before serving.

Goethe's Favourite

Besides the uses listed above, chestnuts can be pickled, candied or made into a purée as a substitute for potatoes.

Chopped, stewed and baked with red cabbage, they make a substantial vegetable pudding—and boiled with Brussels sprouts they were the favourite dish of Goethe, the 18th-century German poet and dramatist.

Walnut

In the Golden Age of Greek lore, when man lived on a handful of acorns, the gods fed upon walnuts and so their name was *Jovis Glans*—the nuts of the god Jove (Jupiter)—this has since become *Juglans regia*, meaning the royal nut.

The tree is a native of Asia Minor and the learned Roman, Varro, mentions it existing in his day in Italy (he was born in 116 BC). Its introduction to Britain is usually set down as the 15th-century, and its name walnut is a contraction of the early German walsh-nut. The Saxons knew it as the welsh (meaning strange or foreign) nut.

The walnut has rarely spread beyond those areas in which it was planted and so is not included in the British flora—but there are enough trees growing wild in hedgerows and around old woodland and parks to justify the nut being included in this 'free food' book.

The walnut is a most handsome tree, growing up to 18 m (60 ft.) in height, with a huge spreading head, twisted branches and grey bark (smooth on a young tree, furrowed on older ones which live up to 200 years). Its flowering is similar to that of the oak and hazel, the sexes bearing different flowers on the same tree. The nuts are ripe in late October. The leaves, looking rather like those of the ash, are sweet-smelling, and sometimes used for pot-pourri, with beans and spices. The walnuts, in a casing at first green then brown, are ripe in October.

*A woman, a steak and a walnut tree,
The more you beat 'em, the better
   they be.*

This old adage has reference to harvesting walnuts. Evelyn says:

'In Italy they arm the tops of long poles with nails and iron for the purpose of loosening the fruit and believe the beating improves the tree—which I no more believe than I do that discipline would reform a shrew.'

Even today, some gardeners believe that a walnut tree that does not bear can, with good results, be beaten in early March when the sap is rising. A bill-hook is the usual weapon—and the bark of many old walnut trees show today the marks of past treatment. Beating breaks some of the long shoots, so encouraging short, fruiting spurs to grow.

Walnuts

Walnuts are best when fairly ripe and dry. If picked before October and November, the young 'wet' walnuts are rather tasteless—the only reason for picking early (in July whilst still green) is to make pickle.

Storing Walnuts

Gather the nuts when fully ripe and falling from the tree—and spread out to dry for a few days in a shed. After that, remove every bit of the husk with a stiff-bristled brush and water—or place the nuts in a tub with sand and water and scrub with a hard broom. Again spread out to dry, on trays in a good current of air.

When the nuts are thoroughly dry, store them in sand in a box. Put a layer of sand at the bottom of your box, then a layer of nuts—adding a sprinkling of salt. Just cover each layer with sand and repeat the process until the box is full. The salt to some degree prevents the walnuts from shrivelling, but even so do not try to keep the nuts too long.

Pickled Walnuts

It is not necessary, as many cookery books tell you, to prick the fresh green nuts all over with a pin before pickling —a task which takes many hours and stains one's hands dark brown (a difficult stain to remove, by the way).

Place the July-picked walnuts in a deep jar and cover with cold brine made of salt boiled with water (about 6 oz. of salt to a quart). Change the brine every four days for 12 days and then leave for nine days—after which you should spread the nuts out on a large dish and leave to dry and go black.

Next, place the dry nuts in smaller jars and pour *spiced* vinegar over them. Leave for three days, then drain. Reboil the vinegar and again pour over the nuts. Finally, cover and keep in a dry cupboard for at least a month before eating.

The necessary spiced vinegar is

made from half a gallon of vinegar to which are added three onions, a pinch of cloves, some peppercorns, a little ginger and a few bay leaves. Boil, strain and then use on the walnuts.

Walnut Fudge

A Cheshire recipe: Put 1 lb. of sugar, $\frac{1}{2}$ pint of cream and a tablespoonful of syrup into a saucepan. Stir well until it comes to the boil, and boil for 20 minutes. At this point stir in $\frac{1}{4}$ lb. of chopped walnuts and half a teaspoonful of vanilla essence—and place the saucepan in cold water.

Stir quickly until the toffee is thick, then empty into a buttered tin, mark into squares with a knife, and leave till cold.

Walnut Scones

A Lanarkshire recipe: Make a light scone mixture and roll out quickly. Sprinkle a teaspoonful of cinnamon, a tablespoonful of sugar and two table-spoonsful of chopped walnuts over the scone mixture and fold in three. Roll lightly to the required thickness, cut into shapes and bake in a quick oven.

Date and Walnut Cake

A Northumberland recipe: Cream $\frac{1}{2}$ oz. of butter and the same amount of sugar, add one egg, 12 oz. of flour, 1 lb. of dates cut small and 4 oz. of walnuts.

Mix with a teacupful of milk containing a teaspoonful of bicarbonate of soda, beat well and then mix in a teaspoonful of baking powder. Bake in a flat, well-greased tin for $1\frac{1}{2}$ hours.

Catchup of Walnuts (1773 recipe)

'Put the peel of nine Seville oranges to three pints of the best white wine vinegar. Let it stand for three or four months. Pound 200 of walnuts just before they are fit for pickling. Squeeze out two quarts of juice, put into the vinegar. Tie a quarter of an ounce of cloves, the same of mace, a quarter of a pound of eschelot [i.e. shallot] in a muslin rag. Put in the liquor. In three weeks boil it gently till near half is consumed. When cold, bottle it.'

Compote of Walnuts

This is a Midlands hunting country recipe from *The Pytchley Book of Refined Cookery* (1886):

'Prepare some new walnuts in the following manner. Split the shells in half, take out the kernels with a knife and peel them without breaking them. Put them in iced or very cold water with some lemon juice. Just before serving drain them, arrange in a glass dish in an oval form and pour over them a reduced syrup with a whip of cream flavoured with any liqueur.'

Walnut oil is highly valued in France as the main ingredient of salad dressings in the walnut-growing areas. To make this oil, pick ripe walnuts and process them in the same way as beechnuts (see next page) but without roasting them.

Beech

Although small, the beech nut is remarkably high in protein and oil and makes a valuable food for humans and some animals.

If the oak is called the Monarch of the Forest, the beech (*Fagus sylvatica*) is entitled to be called the Mother of the Forest—it is justifiably said that

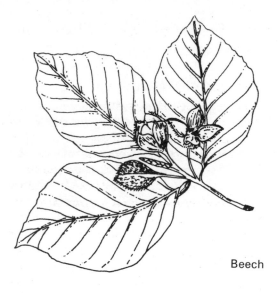

Beech

without beech there could no more be properly tended forests of broad-leaved species, since the growth of many other valuable timber trees is 'nursed' by the presence of beech.

Like the oak, the stately beech needs

no identifying description in this book —and it is common throughout the country, especially on chalk soils. The name Fagus originates from a Greek word meaning to eat but this probably referred to the beechmast (the term for the nuts) as a food for pigs. But beechmast—raw or roasted and salted —is also good for us to eat, tasting like young walnut and slightly bitter.

Beech trees fruit only once every three or four years, but each tree produces a vast amount of beechmast so it is very easy to find enough— although you will have to gather before the squirrels, dormice, badgers and deer have taken it all, and before the beechmast dries out.

Beechnut Oil

Beech nuts can yield up to 20 per cent of their own volume of an oil rich in protein, vitamins and minerals. It makes a distinctive salad dressing and can be used for frying or other cooking, even as a substitute for butter.

Remove any husk, earth and leaves, and roast the beechmast in a moderately hot oven. Grind the nuts through a mincing machine, or in an electric liquidiser. Put the nut pulp in a muslin bag, place it in a strong sieve and press down with a heavy weight.

Oak

There is no need to describe the botanical characteristics of England's traditional tree, or where to find it. Everyone knows, too, what a long-lived tree the oak is—but not so many know that the oak (*Quercus robur*) does not produce an acorn until 60 or 70 years old or that its timber is not ideal for use until the oak is at least 150 years old!

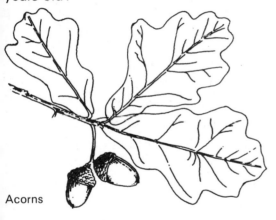

Acorns

In times of famine, acorns, usually considered as pig fodder, have often been used as a supplement to bread flour and, during the last war, as a substitute for coffee. They are also sometimes used as hen food.

Worship of the oak is centuries old. There is no doubt of an oak cult in Indo-Germanic religion, of the connection of the oak with the god of thunder (whether Zeus or Thor), and various sacred associations. Of one very ancient oak in Staffordshire, John Evelyn declared:

Upon oath of a bastard's being begotten within reach of the shade of its boughs—which I can assure you at the rising and declining of the sun is very ample—the offence was not obnoxious to the censure of either ecclesiastical or civil magistrate.

Pine Nuts

These often appear in early recipes. They were probably the small seeds from the pine cones, used as a spice.

Pine

They have a pungent turpentine flavour. Also, the sticky pink aromatic ends of the common fir used to be cooked in a mixture till it was 'charged with the flavour y nough'. The 'fir end' was then taken out (as one withdraws a bay leaf or vanilla pod), and the pine flavouring would be accentuated by the small pine seeds strewn on top as a garnish.

Pokerounce

From old cookery books this seems to have been the term used for a hot, spiced toast, spread with pine nuts.

'Take hony and caste it in a potte till it wexe chargeaunt y now. Take gyngere and canel and galyngale and caste thereto. Take whyte brede and kyt into trencherous and toast and take the paste [i.e. boiled and flavoured honey] whyle hot and spread it upon them with a spoine and plante it with pynes and serve.'

Modern Recipe

Take a jar of honey with a large sticky pink end from a fir branch submerged in it, cover, and put on the stove to bake slowly. When the honey is strongly flavoured, add a pinch of mixed sweet spice and spread it hot on hot buttered toast garnished with pine kernels.